6-00
E-p
$25⁰⁰
VER

W9-CZY-928

Conserving the environment /
GE 300 .C63 1999                                    9141

WRIGHT LIBRARY

# the Environment

Other Books in the Current Controversies Series:

The Abortion Controversy
Alcoholism
Assisted Suicide
Computers and Society
Crime
The Disabled
Drug Trafficking
Ethics
Europe
Family Violence
Free Speech
Gambling
Garbage and Waste
Gay Rights
Genetics and Intelligence
Gun Control
Hate Crimes
Hunger
Illegal Drugs
Illegal Immigration

The Information Highway
Interventionism
Iraq
Marriage and Divorce
Mental Health
Minorities
Nationalism and Ethnic Conflict
Native American Rights
Police Brutality
Politicians and Ethics
Pollution
Racism
Reproductive Technologies
Sexual Harassment
Smoking
Teen Addiction
Urban Terrorism
Violence Against Women
Violence in the Media
Women in the Military

# Conserving
# the Environment

**Laura K. Egendorf**, *Book Editor*

**David Bender**, *Publisher*
**Bruno Leone**, *Executive Editor*

**Bonnie Szumski**, *Editorial Director*
**Brenda Stalcup**, *Managing Editor*
**Scott Barbour**, *Senior Editor*

CURRENT CONTROVERSIES

No part of this book may be reproduced or used in any form or by any means, electrical, mechanical, or otherwise, including, but not limited to, photocopy, recording, or any information storage and retrieval system, without prior written permission from the publisher.

Library of Congress Cataloging-in-Publication Data

Conserving the environment / Laura K. Egendorf, book editor.
      p. cm. — (Current controversies)
   Includes bibliographical references and index.
   ISBN 1-56510-950-3 (pbk. : alk. paper). — ISBN 1-56510-951-1
(lib. : alk. paper)
   1. Environmental management.   2. Environmental protection.
3. Biodiversity.   I. Egendorf, Laura K., 1973–   .  II. Series.
GE300.C63   1999
333.7'2—dc21                                   98-35009
                                                  CIP

© 1999 by Greenhaven Press, Inc., PO Box 289009, San Diego, CA 92198-9009
Printed in the U.S.A.

Every effort has been made to trace the owners of copyrighted material.

# Contents

chemicals on lakes and human health should concern people. This inaccurate media coverage poses a threat to the success of environmental regulations.

## No: There Is Not an Environmental Crisis

# Chapter 2: Should Biodiversity Be Preserved?

## Yes: Biodiversity Should Be Preserved

## No: Humans Should Not Attempt to Preserve Biodiversity

# Chapter 3: How Can Pollution Be Reduced?

## Chapter 4: Can Free-Market Approaches Protect the Environment?

### Yes: Free-Market Solutions Are Effective

### No: Free-Market Approaches Are Not Adequate

lowances, is an ineffective environmental policy. Instead of reducing pollution, emissions trading enables a handful of corporations to purchase the majority of allowances, permitting them to continue to pollute instead of using new pollution control technologies.

    The environmental crisis in America is exacerbated by overconsumption. Purchasing and recycling environmentally safe products will not protect the environment because those practices do not reduce consumption. In order to conserve the environment, Americans need to decrease their reliance on material goods.

    The deterioration of the environment is due largely to the belief that natural resources should have market values. Belief in a market economy leads to environmental degradation and exploitation of the poor. The solution to the crisis is to give local communities control over their resources.

# Foreword

By definition, controversies are "discussions of questions in which opposing opinions clash" (Webster's Twentieth Century Dictionary Unabridged). Few would deny that controversies are a pervasive part of the human condition and exist on virtually every level of human enterprise. Controversies transpire between individuals and among groups, within nations and between nations. Controversies supply the grist necessary for progress by providing challenges and challengers to the status quo. They also create atmospheres where strife and warfare can flourish. A world without controversies would be a peaceful world; but it also would be, by and large, static and prosaic.

## The Series' Purpose

The purpose of the Current Controversies series is to explore many of the social, political, and economic controversies dominating the national and international scenes today. Titles selected for inclusion in the series are highly focused and specific. For example, from the larger category of criminal justice, Current Controversies deals with specific topics such as police brutality, gun control, white collar crime, and others. The debates in Current Controversies also are presented in a useful, timeless fashion. Articles and book excerpts included in each title are selected if they contribute valuable, long-range ideas to the overall debate. And wherever possible, current information is enhanced with historical documents and other relevant materials. Thus, while individual titles are current in focus, every effort is made to ensure that they will not become quickly outdated. Books in the Current Controversies series will remain important resources for librarians, teachers, and students for many years.

In addition to keeping the titles focused and specific, great care is taken in the editorial format of each book in the series. Book introductions and chapter prefaces are offered to provide background material for readers. Chapters are organized around several key questions that are answered with diverse opinions representing all points on the political spectrum. Materials in each chapter include opinions in which authors clearly disagree as well as alternative opinions in which authors may agree on a broader issue but disagree on the possible solutions. In this way, the content of each volume in Current Controversies mirrors the mosaic of opinions encountered in society. Readers will quickly realize that there are many viable answers to these complex issues. By questioning each author's conclusions, students and casual readers can begin to develop the critical thinking skills so important to evaluating opinionated material.

Current Controversies is also ideal for controlled research. Each anthology in the series is composed of primary sources taken from a wide gamut of informational categories including periodicals, newspapers, books, United States and foreign government documents, and the publications of private and public organizations. Readers will find factual support for reports, debates, and research papers covering all areas of important issues. In addition, an annotated table of contents, an index, a book and periodical bibliography, and a list of organizations to contact are included in each book to expedite further research.

Perhaps more than ever before in history, people are confronted with diverse and contradictory information. During the Persian Gulf War, for example, the public was not only treated to minute-to-minute coverage of the war, it was also inundated with critiques of the coverage and countless analyses of the factors motivating U.S. involvement. Being able to sort through the plethora of opinions accompanying today's major issues, and to draw one's own conclusions, can be a complicated and frustrating struggle. It is the editors' hope that Current Controversies will help readers with this struggle.

Greenhaven Press anthologies primarily consist of previously published material taken from a variety of sources, including periodicals, books, scholarly journals, newspapers, government documents, and position papers from private and public organizations. These original sources are often edited for length and to ensure their accessibility for a young adult audience. The anthology editors also change the original titles of these works in order to clearly present the main thesis of each viewpoint and to explicitly indicate the opinion presented in the viewpoint. These alterations are made in consideration of both the reading and comprehension levels of a young adult audience. Every effort is made to ensure that Greenhaven Press accurately reflects the original intent of the authors included in this anthology.

*"International agreement on environmental issues is often difficult to achieve because countries are not at equivalent stages of social and economic development."*

# Introduction

The condition of the environment is a worldwide issue. Air and water pollution do not recognize borders; poor soil conditions in one nation may reduce another country's food supply. At the same time, different regions do face different problems. One key distinction is between the environmental threats faced by developed nations, such as the United States and western European countries, and developing nations, such as India and Mexico. Most agree that these nations may have dissimilar crises, but debate remains over whether the solutions to their problems are unique as well.

The environmental problems faced by developed nations are largely the result of their economic strength and higher standards of living. Overconsumption is cited by many observers as a cause of resource depletion in the First World. Americans, and to a lesser extent western Europeans, Japanese, and other residents of developed nations, are more likely to own one or more cars, purchase more food and clothes than subsistence levels require, and use considerable amounts of electricity. Americans consume a disproportionate amount of the planet's resources. The United States is home to 5 percent of the world's population but uses 25 percent of its resources. Overall, the developed world has 23 percent of Earth's population but consumes two-thirds of the resources. Environmentalists contend that this high level of consumption will ultimately lead to the depletion of the planet's resources, resulting in adverse consequences for human populations. Developed nations have reduced their rate of population growth, so overpopulation is not as great a problem as it was previously considered to be; however, because of the high level of consumption, each new person in a developed nation will use three times as much water and ten times as much energy as a child born in a developing country. The industries needed to create products for consumption also affect the environment through the emission of greenhouse gases and other wastes.

In contrast, the environmental crises faced by developing nations are the result of poverty. For example, Third World countries often lack the resources and sanitation facilities to provide the public with clean water. Tropical deforestation, caused by the slash-and-burn techniques of poor farmers, is another dilemma. However, as Rice University president Malcolm Gillis has observed, farming is not the sole cause of deforestation by poor people: "Slash-and-burn agriculture is not the only manifestation of the effects of poverty on deforestation. In most, but not all, poor nations, the role of poverty in deforestation is magnified by the ever-more-desperate search for fuelwood by impoverished people." This search for wood is exacerbated by the key environmental problem in developing nations—overpopulation. Third World nations may consume vastly less than America and Europe but their population growth rates are much higher. These nations lack the natural resources and social services that will be needed in order to provide their burgeoning populations with adequate food, shelter, and employment in the coming years. As developing nations move closer to First World status, the accompanying growth in industry could also affect the environment, especially through the emission of greenhouse gases. The global warming agreement reached in Kyoto, Japan, in December 1997 exempted developing nations such as China, India, and Mexico from requirements to reduce their emissions. But according to the United Nations, countries exempted from the agreement will create 76 percent of total greenhouse gas emissions over the next 50 years.

The exemptions in the Kyoto agreement (which must be approved by 55 nations but as of this writing has not been submitted to the U.S. Senate for ratification) raise the question of whether developed and developing nations should utilize the same methods in order to conserve the environment. If the environment truly is a worldwide issue, then the solutions may also be universal. However, international agreement on environmental issues is often difficult to achieve because countries are not at equivalent stages of social and economic development.

Developed nations rely significantly on government regulations to protect and restore the environment; however, many analysts—particularly Americans—believe that the same economic forces that create the wealth of developed countries can solve their environmental troubles. Industry, capitalism, and the free-market system might create overconsumption, but they can also solve its ill effects, these commentators maintain. John Hood, the president of the John Locke Foundation, a policy institute that advocates the free market and limited government, writes, "Corporate America's unique contribution to solving real environmental problems will come from innovation—finding new ways to produce goods and services, package and deliver them to consumers, and dispose of or recycle the wastes generated by their own production or by consumption." In contrast, a system in which the government owns all the land or imposes strict command-and-control regulations on people and businesses is seen as ineffective. The poor environmental condition of communist nations is often cited by these observers as evidence of the inability of government regulations to conserve the environment.

As developing nations grow and become more economically self-sufficient, industrial solutions may become more viable in those countries. However, many commentators assert that Third World and post-communist countries should not follow the United States' lead. These observers see industry as the planet's foe rather than its savior; they believe companies are more likely to be motivated by the quest for profit than a desire to preserve the environment. A better way to improve the environment is to rely on a country's indigenous values, many people maintain. For example, some environmentalists believe that the religious traditions of India promote ecologically friendly values, including vegetarianism and a moderate use of resources. They also prefer traditional agricultural methods, which do not rely on pesticides and chemical fertilizers and therefore do not cause groundwater pollution. Frances Cairncross, a senior editor at the *Economist*, is among those who argue that if industry is to be relied upon, it should be as environmentally advanced as possible: "Industry in the developing countries has a special opportunity. Because it is making new, 'greenfield' investments [investing in undeveloped and often unpolluted land], it can leap a stage and go straight to the best modern practice."

As noted earlier, the Kyoto global warming agreement reveals the difficulty of finding universal solutions to environmental problems. Developing nations would not consider even voluntary participation in emission reduction, arguing that such measures would impede their efforts to improve their economies and industries. Even within developed nations, the response to the treaty has varied. In June 1998, the European Union reached an agreement that will reduce their greenhouse gas emissions by 8 percent. However, many people in the United States have more negative attitudes toward the agreement; they assert that achieving the reduced emission levels could hurt the nation's economy. For example, some American analysts contend, companies might move their plants to developing nations, causing job losses in the United States. Moreover, they argue, emission controls could cause U.S. oil and gas prices to rise. Although the Clinton administration played a key role in reaching an agreement in Kyoto, President Bill Clinton is among those who believe developing countries need to limit their own greenhouse gases before the United States can ratify the treaty. Without the participation of the United States—the world's leading polluter—the treaty might not succeed.

As the Kyoto controversy suggests, international agreement over solutions to global environmental problems is not easily attained. The debate over environmental issues in the United States is also divisive. These global and national debates are the subject of *Conserving the Environment: Current Controversies*. In this book, the authors examine such topics as the state of the environment, the preservation of biodiversity, methods for reducing pollution, and whether the free-market system can solve environmental problems.

# Chapter 1

# Is There an Environmental Crisis?

CURRENT CONTROVERSIES

# Chapter Preface

In 1980, Julian Simon, then a professor at the University of Illinois, placed a bet with Stanford University professor Paul Ehrlich and two of Ehrlich's colleagues. The wager involved the future price and availability of five metals. Simon bet that the price of these metals (adjusted for inflation) would be lower in 1990 than they were in 1980, indicating that they were not scarce resources. Simon won the bet.

The wager was one of the most publicized examples of the opposing views Simon and Ehrlich had of the environment and the future of the planet. Simon had long argued that the environment and living conditions were improving and would continue to do so. According to Simon, human intelligence and ingenuity were the keys to improving the environment. He wrote: "Human beings are not just more mouths to feed, but are productive and inventive minds that help find creative solutions to man's problems."

In contrast, Ehrlich has written for over thirty years on what he sees as the danger of overpopulation and the depletion of natural resources. For example, he contends that the available cropland will not yield enough food for the projected future population. Ehrlich did not interpret the drop in the price of metals as indicative of their abundance. Writing in 1996 about the bet, Ehrlich asserted that the recession of the early 1980s reduced the demand for the metals, thereby driving their prices down.

A second wager, proposed in 1995 by Simon, did not materialize. Simon had offered to bet anyone that any trend concerning material human welfare would improve. However, he declined to bet on the fifteen trends suggested by Ehrlich and a colleague. Simon's death on February 8, 1998, ended the possibility of the second bet taking place.

Although the debate between Julian Simon and Paul Ehrlich has ended, the question of whether there is an environmental crisis remains. In the following chapter, the authors debate the state of the environment.

# The Environment Is Deteriorating

## by Lester Brown

**About the author:** *Lester Brown is the president of the Worldwatch Institute, an environmental research organization that publishes an annual* State of the World *anthology.*

Despite all the promises made at the 1992 Earth Summit in Rio de Janeiro [a gathering of world leaders and environmental experts that focused on environmental conservation], the world has added 450 million people, the climate is changing and deforestation is even more serious a problem, especially in the Northern Hemisphere. Canada is losing a million hectares of forest a year, and so is Russia.

In spite of pledges to return to 1990 levels, carbon emissions have drastically increased—altering the earth's climate. The U.S. is up 6%, Japan up 9%. Germany has reduced emissions by 10%, primarily by ending the industrial inefficiency in Eastern Germany. Russia has reduced its emissions by 28%, but economic production is down by 40%. Other problems are species loss, soil erosion, overfishing, water pollution, etc. Poverty is a serious issue as over 1.3 billion people are trying to live on $1 a day or less.

Eight countries together include 56% of the world's population, 59% of its economic output, 53% of world's forested area, and 58% of the carbon emissions. We call these environmental heavyweights the E-8—they include Brazil, China, Germany, India, Indonesia, Japan, Russia, and the United States. We suggest the meetings of the major industrial nations, the so-called G-7 countries, which include only a small fraction of the world's population, become E-8 meetings.

## A Success Story

The biggest success in the international environment has been the reduction in chlorofluorocarbon (CFC) emissions. [The Worldwatch Institute's] *State of the World 1997* looked at this as a case study in what we can learn about international environmental cooperation. It began in 1974, when two scientists at the

Reprinted, by permission, from Lester Brown, "The State of the World and the Region Five Years After the Earth Summit," in *State of the World,* the Southern California Council on Environment and Development (SCCED), March 22, 1997.

University of California at Irvine published a paper in *Nature* that hypothesized that CFCs would set in place a chain of chemical reactions that would deplete the stratospheric ozone layer so essential to shield us from the sun's deadly ultraviolet (UV) radiation. Public and media reaction led to bans on the use of CFCs in aerosol cans in the U.S. and Canada by the end of the decade and in Europe soon after.

In 1977, 33 national governments and the European Community created a World Plan of Action on the Ozone Layer, including global scientific collaborative research. As a result, in 1985 two British scientists reported a hole, larger than the U.S., in the ozone layer over Antarctica that let in unprecedented amounts of UV radiation. It is interesting that NASA had a polar orbiting satellite that had been picking up the problem for years, but the satellite was programmed to ignore such extreme numbers as "errors."

Immediately the National Science Foundation sent a team to Antarctica and decided the hole was caused by CFCs. They held a press conference to announce their tentative findings, because the scientists were fearful that the hole might expand rapidly. The result was the signing in 1987 of the Montreal Protocol to cut production of CFCs in half by 1998. But in 1988 the world's largest producer of CFCs, DuPont, announced it was completely ending production of CFCs before the year 2000.

> *"The 11 warmest years, since record keeping started in 1866, have all been since 1979."*

In 1991, NASA announced that ozone depletion was progressing twice as fast as expected over parts of the Northern Hemisphere, causing a projected additional 200,000 deaths from skin cancer in the U.S. over the next 50 years. At the 1992 Copenhagen meeting, the phase-out date for industrial countries was advanced to 1996 for CFCs and some other ozone-destroying chemicals.

The results in 1996 are a reduction of CFC production by 76% and a projected nearly complete phase-out in the next 10 years. The problem is that some developing countries are continuing CFC production, notably China, India and the Philippines (some of which is now being illegally smuggled back into the U.S. for auto air-conditioners).

## Climate Problems

Our conclusion is that the world can change behavior in response to new information, if we are sufficiently scared about the implications. The purpose of the Worldwatch Institute is to provide information to aid the process of change to move the world toward a sustainable economic system. The question is what might be the climate equivalent of the hole in the ozone layer?

Although governments promised in Rio to reduce carbon emissions to 1990 levels, the results have not been encouraging, as only a few political and corporate leaders are sensing the need to engage environment issues. I have talked

VRJC LIBRARY

about the State of the World in eight European countries and met with three heads of state. They were concerned about the declining ocean fish catch, the scarcity of fresh water, and the extreme climate events.

The recent spate of storms, heat, floods, and droughts are drastically affecting the insurance industry. In 1996, the CEOs of the 60 largest global insurance companies signed a statement urging governments to cut carbon emissions to reduce global climate change. This is a very significant development, but we don't yet have

> *"We are rapidly losing cropland through residential and commercial development and soil erosion."*

critical mass for a major development at the Kyoto Conference in December 1997. [An agreement was reached to reduce the emissions of greenhouse gases. However, the U.S. Senate has not ratified the treaty.]

The 11 warmest years, since record keeping started in 1866, have all been since 1979. In the past century through the 1980s, there were $16 billion in storm-related losses. In the first six years of the 1990s, storms have cost $66 billion, which is causing alarm throughout the insurance industry. Hurricane Andrew took down 60,000 buildings and several insurance companies. In addition, typhoons did $26 billion in damage in China in 1996, forcing 2 million people from their homes.

Global warming is a serious matter, but what will wake us up? What will get the globe moving on an environmentally sustainable path? Falling sperm counts could scare us, or perhaps a new disease from the tropics? Or a dramatic climatic event striking a major metropolitan area? If Hurricane Andrew had hit Miami and New Orleans, it could have tripled the amount of damage.

## Food and Water Shortages

The first economic indicators that will signal that we are on an unsustainable path will be rising food prices. The ocean catch had risen from 18 million tons in 1950 to 89 million tons in 1989, but has stayed the same since 1989. As world population increases, we will see a halving of the ocean catch per capita. Fish farming can't help much because it depends on growing more land-based food to feed the fish. And we are rapidly losing cropland through residential and commercial development and soil erosion.

Water tables are falling in all the world's food growing regions from over-pumping, but water scarcity is hard to photograph for the evening news. In the U.S. farmers greatly expanded irrigation by pumping from the great Ogallala Aquifer. But this is essentially a fossil aquifer, which is not significantly recharged by rain water. It is becoming so depleted now that in Texas irrigated land has decreased by 11%, forcing many farmers to return to less productive dryland farming.

China faces the most serious water problem. Water tables in North Central

China have fallen 35 meters over the last two decades, putting 100 million people at risk. Now the great water reservoirs around Beijing are only available to supply water for the city, leaving millions of farmers to dryland farming. To import a ton of wheat is saving the 1000 tons of water it took to grow that wheat.

## The Costs of Feeding China

Now we have reached the physiological capacity of crops to use more fertilizer. Farmers in the U.S. and Europe are using less fertilizer than in the 1980s because it cannot boost soil fertility by enough to pay for itself. In addition, farmers now must deal with unpredictable local climate change. In July 1995, when Chicago went to a temperature of 105° for a week, 465 people died, and the corn crop withered.

World grain stocks dropped to the lowest level ever in 1996. The additional 80 million people per year, and the growth in affluence leading to increased meat consumption, is putting increasing pressure on a decreasing grain supply. For example, the Chinese now want to diversify their diets to include eggs, poultry, beef, and beer—they want to live like us in the U.S. To increase from 100 eggs per person per year in 1990 to 200 eggs per year will take 1.3 billion more hens. It will take more grain than all Australia exports to get that. One more bottle of beer per person in China takes 270,000 tons of grain. And China is losing cropland at a record rate from expansion of industry and housing.

China can afford to pay for imported grain from its $40 billion annual trade surplus but no one can grow the grain to supply them. Egg and milk consumption are also rising rapidly in India, putting more pressure on the world's resources. This will cause rising grain prices. If grain prices doubled, it would not affect the U.S. much, but would drastically affect the 1.3 billion people in the world that now live on less than $1/day. If they take to the streets, the resulting instability could affect earnings of corporations, stock markets, the international monetary system, pension fund earnings, etc.

## Finding a Solution

But none of the problems are unmanageable. We need to stabilize both world population and climate, and both could be done with existing technology—if we made them global priorities.

Since Rio, 100 cities have developed their own plans for a sustainable future, but we need broader systemic change. To stabilize the climate, we need renewable energy sources, such as solar and wind. I believe wind is going to be the energy source of the future.

Change is possible. In 1990, we crossed the social threshold in Eastern Europe, and one morning we woke up and the Berlin Wall was down and change was irreversible.

# Capitalism Has Worsened the Environment

by *The People*

**About the author:** The People *is a newspaper published by the Socialist Labor Party.*

Periodically, during the last several decades, groups of scientists engaged in the study of varied phenomena vital to the existence and well-being of the manifold forms of plant and animal life that inhabit the planet earth have been prompted to issue warnings that irreversible damage is being done to the planet's fragile life-support system, and that the consequences for the human race could be catastrophic. For example, in the 1970s, the British magazine, *Ecologist,* published an article entitled, "Blueprint for Survival," bluntly warning of a possible worldwide environmental catastrophe "if [then] current trends were allowed to persist."

## Unheeded Warnings

Within days after publication of the "Blueprint" article, a group of 187 British scientists wrote a letter to *The Times* of London, which, though critical of some aspects of the "Blueprint" article, agreed that there was indeed a "growing ecological crisis," adding that there was "now no escape from the necessity for a fundamental rethinking of all our working assumptions about human development in relation to the world we live in."

Barely a month later came a third warning, this time from the Club of Rome, an international organization of scientists and intellectuals, suggesting the need for a "system" that would stabilize population and industrial capacity. This, in turn, was followed by the first of three reports issued by the president's Commission on Population Growth warning that population growth must be slowed and calling upon the United States to set an example for the world by becoming "the first nation to adopt a deliberate population policy. . . ." (*The New York Times,* March 12, 1972.)

Despite such warnings, despite some efforts to promote conservation and re-

Reprinted by permission from *The People*, "'State of the World': Environmental Dangers Threaten Human Survival," *The People*, March 26, 1994.

cycling programs, despite attempts to reclaim some waterways and clean up some toxic wastes, despite sporadic attempts at reclaiming or replenishing some lands, environmental deterioration has continued apace. Now, we are again being bombarded with warnings of potentially dire consequences for humanity inherent in a growing imbalance between the needs of an increasing world population and the continuing deterioration of the environmental and ecological resources required to sustain it.

## Food Supply Loss

In the summer of 1993, Worldwatch Institute, a research group, issued a warning, echoed by United Nation experts, that the ongoing degradation of the world's crop lands and the ongoing depletion of the world's fisheries were not merely slowing, they were in important respects reversing, the growth of food supplies. Worsening the problem, the group said, was an ominous decline in the effectiveness of fertilizers.

Several months later, the U.N. Food & Agriculture Organization reported that as a result of what it called "industrial agriculture" policies, market incentives and overexploitation, world crop diversity was being lost, domestic breeds of animals were being depleted and the biological diversity of aquatic species was being threatened. Collectively, such loss of diversity, in the words of the U.N. organization's director-general, "threatens world food production and could eventually put human beings at the top of the endangered species."

*"Despite some efforts to promote conservation and recycling programs, . . . environmental deterioration has continued apace."*

The warnings didn't stop there. In mid-January 1994, Worldwatch Institute issued its "State of the World 1994" report. It painted a grim picture of the rapidity with which "human demands are approaching the limits of oceanic fisheries to supply fish, of range lands to support livestock, and in many countries, of the hydrological cycle to produce fresh water." As Worldwatch president, Lester Brown, summed it up in the report's final chapter, "The bottom line is that the world's farmers can no longer be counted on to feed the projected additions to our numbers."

## The Population Issue

The report also focused attention on the need for the 1994 World Conference on Population and Development to reach an understanding of how many more people the planet can support. Presumably that understanding would—or should—lead to agreement on practical, common-sense decisions for dealing with the threat.

(Parenthetically, it ought to be noted here that—whether intended or not—the repeated emphasis on population growth tends to distort the nature of the immediate problem. For whatever problems a steadily growing world population might eventually present, the current threat stems not from overpopulation, or even excessive expansion, etc., but from the past and ongoing devastating effects of the obsessive ruling-class drive for profits with an utter disregard for the damage that is being done in the process to the earth's life-sustaining ecological and environmental systems.)

All experience and present indications, however, hold out little hope that such will prove to be the case. For example, when the president's Council on Sustainable Development considered the growing problem at a meeting it held in January 1994 in Seattle, it saw the essence of the problem as one that is "dramatized in the battle between jobs and the environment." (*San Jose Mercury News,* Jan. 16, 1994.) Moreover, as the *Mercury News* item noted, "the theme of the Seattle meeting, and of 1994's Worldwatch study, is to find ways to expand the economy without damaging the environment." What that means in practical terms in the current economic, political and social environment is "business as usual."

## Capitalism Is the Problem

Accordingly, among the issues specifically mentioned in media reports as requiring "close attention" are "forest management, energy production, transportation and water," all of which are capitalist problem areas, no doubt. Indeed, the "State of the World 1994" already deals with a host of such issues. However, given the profit-oriented nature of capitalist society and the fiercely competitive battle for markets that it engenders, there is little reason to hope that those problems were, or are to be, the subject of new and revolutionary approaches aimed at the basic cause of the growing threat.

On the contrary, all indications are that the issues were and will continue to be "considered" in an environment that assumes and accepts the inevitability of capitalism. Consequently, regardless of the to be expected moral and ethical preachments about the need to recognize the threats and to take vaguely generalized steps to deal with them, the perspectives, discussions and proposals will be decidedly limited to concepts that can only serve to perpetuate the evil social system that is primarily responsible for every one of the threats to the welfare of humanity.

Ignored, as usual, will be the fact that the production of food, like the production of any other commodity, is governed basically by the narrow, antisocial demands of private interests. Neither is there any convincing indication that the greed, avarice and

> *"The harm and damage already done to humanity by the continued existence of capitalism . . . is beyond exact calculation."*

wanton exploitation of the land or the reckless depletion and/or destruction of the natural resources that mark the frenzied pursuit of profit by the ruling classes of the world will be recognized as inherent in the prevailing capitalist system. In short, once again, no real or pointed effort will be made to place the blame for the growing problems where it belongs, on the real culprits, the capitalist system and its profit-driven capitalist class.

## Socialism Is the Solution

The evidence is overwhelming that no social problem can be solved or social threat eliminated as long as its cause is left untouched. Accordingly, it is the subjugation of food production, as well as of environmental and ecological matters, to the interests of profit considerations that create the issues that keep frustrating all efforts at solutions. Only the naive can still believe that the greed and irrational pursuit of markets and profits that capitalism inherently generates can somehow be harnessed and the social, political and economic activities of the capitalist system brought into harmony with the ethical, moral and humane approaches that the solution to our manifold problems demands.

The issue confronting us, literally, is survival. The harm and damage already done to humanity by the continued existence of capitalism far past its progressive evolutionary stage is beyond exact calculation. With each passing day it gets worse. For the fact is that nothing constructive can be done about the growing world population or the depletion of our natural resources, or about any of the other serious ills afflicting society today, until socialism has been established.

# Global Warming Is a Serious Problem

**by Darren Goetze**

**About the author:** *Darren Goetze is a staff scientist for the Union of Concerned Scientists, a nonprofit alliance of scientists and citizens that combines research, advocacy, and education in an effort to ensure a safer environment.*

Global warming is no game. The risks to human beings—to our health, our food supply, and our housing—are great. And the risk to the animals and plants with which we share the earth is also enormous. We are standing at a crossroads. Continuing on our current path will lead us down the road to ruin. But we don't necessarily have to choose that path. We can choose a different road and a different destination if we make the right choices soon.

## Global Warming Does Exist

*Evidence.* In 1995, the Intergovernmental Panel on Climate Change, an international body of climate scientists brought together by the United Nations, concluded that global warming is already a serious problem. The average surface temperature of the earth has already increased by 0.5° to 1.1° Fahrenheit (0.3° to 0.6° Centigrade) since the last half of the 19th century. And the climate has already begun to change: all of the 10 warmest years on record have occurred since 1982.

Could these changes be due only to the climate's natural variability? The Intergovernmental Panel thinks not. Their *Second Assessment Report* provides evidence that heat-trapping gases related to human activities—such as carbon dioxide from burning coal, oil, and gas—are in part driving global warming by increasing the amount of the sun's heat trapped in the earth's atmosphere. This extra heat is making the global climate system unstable. The Intergovernmental Panel concluded that the rise in temperature and change in climate are "unlikely to be entirely natural in origin" and that "the balance of evidence suggests a discernible human influence on global climate."

Reprinted, by permission of the Union of Concerned Scientists, from Darren Goetze, "Global Warming: The Road to Ruin?" *Nucleus*, Summer 1997.

*Chapter 1*

## The Dangers of Global Warming

*Impacts.* If increases in the atmospheric concentrations of carbon dioxide and other heat-trapping gases go unchecked, more changes in global climate patterns will occur. Because the climate system is complex, scientists cannot predict precisely how much and how fast the climate will change. But sophisticated computer simulations project a range of scenarios for increases in average surface temperature between 1.8° and 6.3°F (1° and 3.5°C) by the year 2100. (Bear in mind that seemingly small changes in temperature can produce major changes in climate. During the last Ice Age, global temperatures were only 5° to 9°F cooler than they are today, but that was sufficient to bury what is now Canada, New York, and New England under a kilometer of ice.)

Global warming will not generally mean more pleasant temperatures. Within the next 20 years, various regions of the world may experience severe changes in climate. Some may be vulnerable to longer droughts, others to more coastal flooding, and many to more frequent bouts of extreme weather. And if global warming continues unchecked, we could well see
  • greater risk to human health as diseases previously found in tropical areas spread to higher latitudes and elevations
  • severe stress on forests from higher temperatures and/or less rainfall
  • decreases in mountain habitats and the plants and animals that live in them
  • expansion of deserts
  • disruption of agriculture through regional changes in temperature and water resources
  • a rise of 6 to 37 inches in sea level, with persistent flooding endangering coastal wetlands and human settlements

*Human Health.* According to a 1996 World Health Organization report, *Climate Change and Human Health*, changes in the global climate could increase the frequency of extreme weather, such as heat waves, floods, and storms. These events heighten the risk of injury, disease, and death, but also threaten health through shortages or interruptions in the supply of food, water, and power. For example, major flooding could contaminate drinking water supplies, even in the United States.

The reduction of fresh water supplies because of changes in regional rain and snowfall may cause a higher incidence of some water- and food-borne diseases and parasites. Also, insect-borne diseases such as malaria

> *"Changes in the global climate could increase the frequency of extreme weather, such as heat waves, floods, and storms."*

and dengue fever are already moving northward from tropical regions. These changes are unlikely to cause large epidemics in the United States because of the high standards of public hygiene and health-care quality. But public health care may require more resources as the number of people affected by these diseases grows. Americans who are poor or have inadequate access to health care may suffer disproportionately.

*Agriculture.* While impacts of global warming on crop yields and productivity will vary considerably by geographic region, several studies suggest that maintaining agricultural productivity will be difficult in many areas. For example, a study by the US Department of Agriculture indicates that soil moisture losses could reduce agricultural opportunities in the Corn Belt and the Southeast. Maintaining agricultural production may require costly adaptation strategies. Farmers may need to change the types of crops they grow (from corn to wheat, for example) and the cultivation methods they use, and increase the amount of land under cultivation. The USDA study found that if corn production were to suffer significantly, livestock production would also suffer, since animal feed would be less available.

> *"Governments and industry must work together to design policies to reduce the emissions of heat-trapping gases."*

Many regions of the world are likely to be far worse off because they lack the economic resources to adapt agriculture to climate changes. Scientists expect agricultural productivity in parts of sub-Saharan Africa, southeast Asia, tropical Latin America, and some Pacific island nations to be highly vulnerable. The poorest people in these regions may be at greater risk for hunger and famine.

*Natural Habitats.* Forests and wetlands provide critical benefits to human health, by filtering our air and water, and to human welfare, by providing opportunities for recreation and commerce. Changes in regional climate put many such ecosystems at risk by hindering their ability to grow and regenerate. For example, changes in temperature and rainfall could shift habitat boundaries dramatically—alpine ecosystems could shift to higher elevations, and forests could be forced northward. Since climatic changes may well occur rapidly, some species may be unable to adapt or migrate. Many could face extinction.

Aquatic habitats, such as coastal wetlands, are also vulnerable to global warming. The survival of these wetlands—often areas of high biodiversity that also provide protection against floods—depends on the water's temperature, flow, and level. Scientists are confident that global warming will reduce the area of wetlands and change their distribution. Arctic and subarctic wetlands, which are critical refuge and breeding grounds for large numbers of migratory species, are among the most vulnerable. Other coastal zone habitats—including marshes, mangroves, coral reefs and atolls, and river deltas—will also be threatened.

## Possible Strategies

*Solutions.* Avoiding these costly damages justifies immediate action to turn off the road to ruin. But the United States shows no sign of such action. It will not meet its 1992 commitment, made at the Rio Earth Summit, to stabilize emissions of heat-trapping gases at 1990 levels by 2000; by 1995, US carbon dioxide emissions were up 5.5 percent over 1990 levels.

But this doesn't mean that emissions can't be reduced. Scientists and economists have identified many technically feasible, cost-effective opportunities for emissions reductions, including energy-efficiency measures, advanced vehicle technologies, cuts in oil and coal subsidies, and investments in clean, renewable energy sources like wind and solar power. These strategies would yield bonus benefits by reducing local air pollution and creating high-technology jobs. But to take advantage of these opportunities, governments and industry must work together to design policies to reduce the emissions of heat-trapping gases. Recently, 2000 prominent economists issued a statement affirming that climate-friendly policies will not harm the US economy and can, in fact, strengthen it. The Union of Concerned Scientists (UCS) is working in various states, in the Capitol, and in the international community to advance policies that will turn our society aside from the ruinous road to a global warming future.

# Environmental Change Poses a Threat to Food Production

by Paul R. Ehrlich and Anne H. Ehrlich

**About the author:** *Paul R. Ehrlich is Bing Professor of Population Studies in the department of biological sciences at Stanford University in Palo Alto, California. Anne H. Ehrlich is an associate director and policy coordinator at Stanford's Center for Conservation Biology.*

Ever since Reverend Thomas Malthus at the end of the eighteenth century warned about the dangers of overpopulation, analysts have been concerned about maintaining a balance between human numbers and the human food supply. That concern remains valid today in a world where a tenth of the population goes to bed hungry each night and millions die every year from hunger-related causes. Few subjects have been closer to our hearts and minds for the past three decades than the race between population growth and increasing food production. That race was a major focus of Paul Ehrlich's first popular book, *The Population Bomb,* published in 1968; it was also the principal focus of our 1995 book, written with our colleague Gretchen Daily, *The Stork and the Plow.*

The world food situation has been a favorite arena for brownlash [people who lead the backlash against pro-environmental policies] writers and spokespeople who deny, often vehemently, that a growing population might someday run into absolute food shortages. The essence of their argument takes two forms: population growth is not a problem and (for some of them) is even virtually an unmitigated blessing; and food production can be increased more or less forever without constraint. Some of the more extreme holders of the latter view still occasionally quote an old and long-discredited estimate publicized by Catholic bishops several years ago that theoretically 40 billion people could be fed on Earth. Needless to say, these groups aren't fond of our positions on these matters, and the brownlash attacks our analyses regularly.

Reprinted, by permission, from Paul R. Ehrlich and Anne H. Ehrlich, *Betrayal of Science and Reason: How Anti-Environmental Rhetoric Threatens Our Future*, Island Press, 1996.

Let's take a look at one of the brownlash claims about population and food. . . .

## False Agricultural Optimism

*Even though millions of people are still inadequately nourished, advances in agriculture will eliminate the remaining pockets of hunger early in the twenty-first century.*

The argument that technology will save us is a frequent theme of the brownlash, here applied to agriculture. The claim is rooted in past technological successes but is usually made without considering the totally unprecedented nature of today's situation. Technological optimism is nowhere more rampant than in connection with increasing food production.

In addition, agricultural optimists often exaggerate past successes by choosing a time scale for comparison that is congenial to the notion that hunger can be easily eliminated. Thus they compare total food production in 1950 directly with that of the present, calling attention to the great increase in output over the intervening period. This is proof, they suggest, that hunger will be easily eliminated. But simply considering the increase over the entire period obscures the much less favorable trend that started in the mid-1980s, as we discuss later in this viewpoint.

The optimists also ignore or understate the depletion of natural capital, biophysical limits beyond which yields simply cannot be increased, and other factors that make a repetition of the 1950–1985 food production surge very unlikely.

> *"Agricultural optimists often exaggerate past successes by choosing a time scale for comparison that is congenial to the notion that hunger can be easily eliminated."*

They do not point out, moreover, that the institutions and infrastructure needed to translate technological developments into greater agricultural productivity are largely absent in food-short (i.e., less developed) regions and that financial support for those institutions has diminished in recent years. Furthermore, they either neglect to mention or dismiss the potential impacts of global change on food production. And, finally, they seem unaware of the principle advanced by the eminent demographer Nathan Keyfitz that "bad policies are widespread and persistent."

## The Threat of Food Shortages

Knowledgeable scientists are greatly concerned that these constraints on food production may soon result in serious food shortages. As Mahabub Hossain, head of the Social Sciences Division of the International Rice Research Institute (the organization that created the green revolution in rice, the grain that sustains more people than any other), stated in 1994:

> The race to avoid a collision between population growth and rice production

in Asia goes on, amid worrying signs that gains of the recent past may be lost over the next few decades. . . . If [current] trends continue, demand for rice in many parts of Asia will outstrip supply within a few years.

Since Hossain's comment appeared in the *International Herald Tribune,* the development of a new, higher-yielding variety of rice was announced, which is expected to increase rice production roughly 10 percent after full field testing and deployment. This was good news following a period when agronomists feared they had encountered a "yield cap" on rice—that is, a biophysical limit beyond which increased seed production by a rice plant would be impossible. But this latest breakthrough is no guarantee that others will follow; on the contrary, there's every reason to think the limits aren't far off. Moreover, a 10 percent increase in rice production may sound impressive, but set against growing human numbers in Asia (where 90 percent of the world's rice is consumed), that increase would be barely sufficient to support five years of population growth.

## Expecting the Unimaginable

We suspect agricultural optimism can be traced mostly to a great faith in the potential of science to pull technological rabbits out of a hat. An example of that optimistic attitude appeared in one of the world's best business journals, *The Economist.* The article discussed the probable impact in China of both population growth and increasing per-capita demand for grain (accompanied by losses of farmland). After describing a pessimistic assessment by agricultural policy analyst Lester R. Brown, president of the Worldwatch Institute, and several Chinese scholars, the last paragraph exemplified this blind faith in technology: "Optimists counter the gloomy Malthusians by pointing to as yet unimagined scientific breakthroughs that will boost crop yields around the world. Economists add the simple point that China could then easily export other goods to pay for its imported grain."

The first statement, converted to a sports analogy, is like claiming that "unimagined breakthroughs in training will allow a runner to run a two-minute mile." The second sentence simply ignores the economics of supply and demand. If China's demand skyrockets, grain prices will be driven up, and global access of people to food will quite likely *decrease.* The agricultural scientists we know don't look to "unimagined breakthroughs" to solve their problems; that would hardly be a prudent strategy. Rather, they are deeply concerned about keeping humanity nourished. As for China's grain imports, what other countries produce enough surplus grain to fill

> *"The agricultural scientists we know . . . are deeply concerned about keeping humanity nourished."*

the gap for over a billion consumers? China's need for imported grain by 2025 could exceed all the grain traded today on the world market.

Analyses of food production trends over the past few decades suggest that there is indeed cause to worry about maintaining food supplies. It is certainly true that the most important indicator of human nutrition, world grain production, has roughly tripled since 1950. What the food optimists overlook is that the rate of increase has markedly slowed since the 1950s and 1960s. More sobering is that since 1985, grain production increases have failed to keep up with population growth, even though population growth itself has slackened; and since 1990 there has been no increase in absolute terms, causing severe shrinkage in grain reserves by 1996.

> *"Rapid climate change could deliver the coup de grace to humanity's chances of even restricting hunger to the present levels."*

There are many reasons for this change. It's true, as Dennis Avery points out, that some of the slowdown can be attributed to changes in agricultural policies and land use in rich countries—enacted in part to reduce surpluses. But far more relevant to future food production are tightening constraints such as degradation and losses of land, limited water supplies, and biophysical barriers to increased yields, all of which are increasingly evident.

## The End of the Green Revolution

The agricultural achievements of the past half-century did not come easily; they took a great deal of effort, cooperation, and investment. Even before the great post–World War II surge in agricultural productivity, the tools needed to do the job were in hand: fertilizer-sensitive crops, synthetic fertilizers, and irrigation technologies. Effecting such far-reaching changes required decades of lead time, and the result is now clear, in terms of both impressive success and growing constraints.

The green revolution [the increase in grain production since the 1950s due to the use of fertilizer, pesticides, and high-yielding seed varieties] has already been put in place in most suitable areas of the world, and most of the expected yield gains have been achieved. Farmers now are seeing diminishing returns from fertilizer applications on some major crops over much of the world, and the biological potential for genetically increased yields in some crops may now be approaching the maximum. Even if some unanticipated breakthrough were to be made, it would take many years, if not decades, to develop and deploy new crop varieties—years during which demand would continue rising as the population expands.

A global study by the United Nations Environment Programme found that significant degradation, ranging from slight to severe, has occurred on vegetated land on every continent just since 1945. This deterioration is the result of human activities—chiefly overcultivation, overgrazing, and deforestation. Soils in many areas have deteriorated beyond the ability of fertilizers to mask the impacts of soil erosion. A subsequent study estimated that the capacity of all the world's productive land to supply food and other direct benefits has declined by about 10 percent. And the degradation continues.

Land is not only being rapidly degraded; it is also increasingly being diverted to uses other than food production. Most commonly, land is taken over simply for living space. As the population expands, more people need room for housing, stores, offices, industry, roads, and other infrastructure. Millions of acres of prime farmland are being lost to urbanization each year.

## Further Threats to Food Production

Impending severe water shortages also threaten world food production. Providing a dependable and abundant supply of water is essential for achieving the high yields of modern grain varieties; indeed, the great increase in grain production of past decades can be ascribed largely to the more than doubling of the amount of irrigated land since 1950. But today, most of the readily available sources of irrigation water have already been tapped, and the trend is beginning to reverse. More and more land is being taken out of production because of the rising costs of pumping water from depleted aquifers or because of the classic problems of irrigation: salting up or waterlogging of soils. In addition, urban demands for water often outbid farmers in water-short areas. By 1980, irrigated acreage was no longer expanding faster than the population, and the day when it ceases to increase at all may not be far off.

More ominous, perhaps, is the possible impact of global climate change on agricultural ecosystems. Rapid climate change could deliver the coup de grace to humanity's chances of even restricting hunger to the present levels. Frightening indeed is the possibility that long-standing climatic patterns could be disrupted by more frequent or severe floods or droughts, farming areas could experience too much or too little rainfall or temperatures higher or lower than those at which crops currently planted thrive, and so on. Such changes could be exceedingly disruptive to farming. They would necessitate adjusting to a new regime, during which time food production would likely drop. And since the climate is unlikely simply to shift to a new, stable regime overnight, farmers may suffer a protracted period of grappling with the vicissitudes of an unpredictable climate.

> *"Even the most optimistic assumptions . . . suggest that at least 2 to 3 billion more people will need to be fed within a few decades."*

Climatologists on the International Panel on Climate Change think that climate-related agricultural problems will be most severe in low-latitude developing nations, where agriculture is less adaptable than it is in richer countries like the United States. We suspect the reverse is true, for several reasons. First, modern intensive agriculture as practiced in rich countries centers on monocultures—large-scale plantings of a single genetic strain of a crop, usually a high-yielding vari-

34

ety. These strains are often more sensitive to adverse weather than are traditional crop strains. Second, the most agriculturally productive regions are in temperate zone areas where exceptionally good soils and climate coincide. Climate change, however, may decouple the favorable climate and the good soils, leaving farmers struggling to maintain food production with new handicaps—either poorer soils or a less benign climate.

In contrast, traditional agriculture in developing nations depends less on monocultures and frequently involves planting several crops together to provide some insurance against failures. We would put our bets on the flexibility of traditional farmers. They not only might adapt more readily to changing conditions than can operators of large industrial farms, they also don't depend on recommendations of an agricultural administration to change crops or planting times. Nor are they constrained by government subsidies and restrictions on their crops. Furthermore, most global warming models project relatively little change in tropical climates from greenhouse gas buildups; more dramatic changes are likely in the temperate zones. In all food-growing regions, global climate change is expected to manifest itself in the short to medium term principally as abnormally severe storms, floods, droughts, and other disruptive weather factors, all of which are potentially catastrophic for agriculture.

## Population Reduction

The impressive array of tools developed in the 1930s and 1940s to expand food production were deployed around the world in the 1960s and 1970s and did their job well. A new kit of tools is required to carry us into the future; yet no such kit appears to be on the horizon, although some help will be provided by genetic engineering. Meanwhile, the world's farmers must persist in their monumental struggle to increase production in the face of deteriorating natural capital and faltering ecosystem services (such as natural pest control being disrupted by misuse of pesticides), and they must win the battle year after year without doing irreparable damage to those services. Continued success might be assured, but only if (among other things) the size of the human population can soon be stabilized and a gradual decline initiated. Yet even the most optimistic assumptions for success in reducing population growth suggest that at least 2 to 3 billion more people will need to be fed within a few decades.

A potentially powerful approach to ending and eventually reversing population growth more rapidly would be greatly increasing socioeconomic equity in opportunity between the sexes and among families, regions and nations—as we have discussed elsewhere. To a large degree, past successes in raising crop yields have come from pushing back biophysical frontiers. Now humanity is faced with pushing back socioeconomic frontiers. That may be the toughest task of all. Confronted with such an enormous challenge, it is difficult to be complacent.

# Some Journalists Understate the Environmental Crisis

## by Ron Nixon

**About the author:** *Ron Nixon is an editor for* Southern Exposure *magazine and a writer on environmental issues.*

"Earth Day alarmists had it wrong," proclaimed conservative columnist Joseph Perkins of the *San Diego Union-Tribune* in a pre–Earth Day commentary (5/1/95).

Drawing on a *New Yorker* article (3/10/95) written by veteran *Newsweek* writer Gregg Easterbrook, Perkins takes to task liberals and "greenies" for "grossly overstating the prospects of global warming, the threat of species extinction and the health risk of pesticides."

### Gregg Easterbrook's Views

"We have heard similar alarmist rhetoric on Earth Day," wrote Perkins, a one-time aide to Dan Quayle. "The warnings of environmental calamity should be greeted with skepticism."

Perkins is hardly alone in his use of Easterbrook's work to denounce the environmental movement. Writing for *Newsweek* (3/10/95), Robert J. Samuelson asserts, "Easterbrook is a true environmentalist. He loves to hike and observe wildlife. But Easterbrook is also an acute reporter offended by ill-informed doomsayers. While environmentalists constantly discover a new 'crisis,' he finds dynamic progress and few impending calamities."

Since the publication of the *New Yorker* article and his recently released book, *A Moment on the Earth: The Coming Age of Environmental Optimism,* described by one critic as "745 pages of feel-good optimism" (*Rachel's Environment and Health Weekly,* 3/13/95), Easterbrook's work has become useful ammunition for media outlets, conservatives, corporate PR departments and

Reprinted, by permission, from Ron Nixon, "Limbaughesque Science: 'Eco-Realism' vs. Eco-Reality," *Extra!,* July/August 1995.

lobbyists against attempts to strengthen the nation's environmental laws.

But in spite of the conservatives' embrace of his work, Easterbrook tries to appear objective in the environment debate and to distance himself from the likes of the late Dixy Lee Ray or Rush Limbaugh, criticizing both liberals and conservatives for what he sees as overzealousness. "The left is afraid of the environmental good news because it undercuts stylish pessimism," he writes in the *New Yorker.* "The right is afraid of the good news because it shows that gov-

> *"Critical examination . . . reveals that many of [Gregg Easterbrook's] conclusions are based on dubious interpretations or even outright distortions."*

ernmental regulations might actually amount to something other than wickedness incarnate, and actually produce benefits at an affordable cost." This pose of centrism has helped win his book wide acceptance among media pundits, who tend to think the truth is always somewhere in the middle.

Yet despite his willingness to give credit to environmentalists, Easterbrook's conclusions are hardly distinguishable from the conservatives he criticizes. "Easterbrook comes across as a sophisticated Rush Limbaugh," writes Peter Montague of the newsletter *Rachel's Environment and Health Weekly* (3/13/95).

Worse, critical examination of Easterbrook's work reveals that many of these conclusions are based on dubious interpretations or even outright distortions. "Easterbrook's labors on his word processor have scant relationship to the real world," Ken Silverstein and Alexander Cockburn write in the newsletter *Counterpunch* (5/1/95).

In a report entitled "A Moment of Truth: Correcting the Scientific Errors in Gregg Easterbrook's *A Moment on the Earth*" (3/18/95), the New York–based Environmental Defense Fund (EDF) points out several examples of Easterbrook's Limbaughesque inaccuracies on the environment

"Although some mistakes are inevitable given a work of this size, Easterbrook's are so numerous and so one-sided in their minimization of the seriousness of environmental problems that they must be addressed," said the authors of the report, which scrutinized four chapters from the book. "Moreover, he has included some assertions in his book after having been warned by technical experts that they were incorrect."

## Distorting Climate Trends

Some of the errors and distortions found by these and other critics:

*Easterbrook:* "In February 1992 the Gallup Organization polled members of the American Geophysical Union and American Meteorological Society, the two professional groups for climatologists. Only 17 percent said warming trends so far convinced them an artificial greenhouse effect was in progress."

*Reality:* This "fact" (which echoes a claim made by *The National Review,*

George Will and Limbaugh) was disavowed by Gallup: "67 percent of those scientists directly involved in global climate research say human-induced warming is now occurring," the polling group said in a press release (*San Francisco Chronicle*, 9/27/92). Only 11 percent said that such warming was not occurring, while the remainder were undecided.

*Easterbrook:* "Scientific support for the notion of a drastic rise in sea level has waned rapidly. . . . The highest observed actual sea-level rise in this century is a mere one inch."

*Reality:* Data from the Intergovernmental Panel on Climate Change (IPCC), one of the most respected scientific authorities on the subject of global warming, show that over the past 100 years, the global average sea level has risen by four to eight inches. This is "not a minor error," EDF notes, "since, for example, this is large enough to have eroded over 40 feet of a typical barrier beach on the East Coast of the United States."

*Easterbrook:* "No research has yet shown that industrial chlorine emissions cause any public health or general ecological harm."

*Reality:* A study by the International Commission on the Great Lakes, a group of U.S. and Canadian scientists, found that chlorine has a devastating effect on the ecological health of the Great Lakes. Other studies have linked chlorine-based chemicals to breast cancer (*Rachel's Environmental and Health Weekly*, 3/13/95).

## Chemical Fallacies

*Easterbrook:* "Today's standard is that chemicals are assumed dangerous, and must be elaborately tested before introduction, with the burden on the manufacturer to establish safety rather than on critics to establish threat."

*Reality:* According to the EPA, of the nearly 70,000 chemicals in commercial use today, only 1 percent have been fully tested for their effects. Seventy-five percent have no testing whatsoever, and 500 new chemicals are introduced each year, many with little testing. Often the burden of proving harm from toxins falls squarely on the shoulders of citizens ("Not Just Prosperity," National Wildlife Federation, 1994).

*Easterbrook:* "In the last decade, environmental litigators have pressured the Fish and Wildlife Service to list creatures at any sign of population decline, regardless of whether the decline appears to engage a threat of extinction. This means a common invocation of doomsday cant—that 'more and more creatures are being listed as endangered every day'—is deceptive, since the listings are based on increasingly lenient criteria and now may be registered even when a creature is numerous."

> *"The success of Easterbrook and others contributes to a climate already hostile to the environment."*

*Reality:* "Most species are listed too late rather than too early to ensure their survival," EDF reports. "According to a recent study, the median population size of an animal species at time of listing was just under 1,000—well below the level generally considered viable. For plant species, the median population size was fewer than 120 individuals, and 39 of these species were listed with ten or fewer known members."

## Corporate Abuse of the Environment

*Easterbrook:* "Environmental initiatives worked well even in their early years, when they were driven by top-heavy federal edicts. They work even better as new regulations have centered on market mechanisms and voluntary choice . . . [such as] a free-market program under which companies trade pollution 'allowances' with each other." (*New Yorker*, 3/10/95)

*Reality:* Pollution trading, instead of reducing pollution, merely transfers the problem from one area to another and promotes inequality. In the first pollution trade under the Clean Air Act in 1992, a company in Wisconsin traded its pollution credits to a company in Tennessee where the minority population was seven times the national average (*The Nation*, 9/10/92).

*Easterbrook:* "Many corporate types . . . are gradually abandoning the notion of adversarial relationships with environmental regulators" and demonstrating "respectful environmental behavior."

*Reality:* The present push in Congress toward radical environmental deregulation is being driven by corporate lobbyists. An *ABC News* investigation (*World News Tonight*, 6/11/95) recently found that many of the proposed changes in environmental regulations were copied directly from an industry lobbyist's memo. One proposal is a non-enforceable agreement to permit industries to dump unlimited amounts of pollution into rivers if they promise to clean them up.

"What we are seeing is finger-to-the-wind journalism," Bud Ward of the monthly newsletter *Environment Writer* said. "There is a very rich mother lode for journalists who are seen as challenging conventional wisdom."

## The Danger of Eco-Realism

Indeed, Easterbrook joins a growing number of journalists who call themselves "eco-realist" or the "sky-is-not-falling movement," which includes John Stossel of *ABC News*, Keith Schneider of the *New York Times*, and Boyce Rensberger of the *Washington Post*. These journalists, working for some of the most powerful news outlets in the country, have managed to construct an image of themselves as iconoclastic outsiders.

Schneider recently left the *Times* to head a Michigan land-use consulting firm, leaving Easterbrook as the most well-known print journalist in the "eco-realist" movement. The success of Easterbrook and others contributes to a climate already hostile to the environment, at a time when proposals are underway to gut the very laws and regulations that have contributed to the success in our environmental health that Easterbrook tells us to celebrate.

"Gregg Easterbrook has been an environmental writer for several years," wrote Peter Montague (*Rachel's Environment and Health Weekly*, 3/13/95). "He has traveled the world, gleaning first-hand information reported in this book. He has read hundreds of scientific studies and he spent five years preparing his manuscript. Yet he feels compelled to support his case by omissions, distortions and fabrications. What a sad waste of a talented writer."

# The Environment Is Improving

by *The Economist*

**About the author:** The Economist *is a British weekly newsmagazine.*

Predictions of ecological doom, including recent ones, have such a terrible track record that people should take them with pinches of salt instead of lapping them up with relish. For reasons of their own, pressure groups, journalists and fame-seekers will no doubt continue to peddle ecological catastrophes at an undiminishing speed. These people, oddly, appear to think that having been invariably wrong in the past makes them more likely to be right in the future. The rest of us might do better to recall, when warned of the next doomsday, what ever became of the last one.

## The Club of Rome's False Predictions

In 1972 the Club of Rome published a highly influential report called "Limits to Growth". To many in the environmental movement, that report still stands as a beacon of sense in the foolish world of economics. But were its predictions borne out?

"Limits to Growth" said total global oil reserves amounted to 550 billion barrels. "We could use up all of the proven reserves of oil in the entire world by the end of the next decade," said President Jimmy Carter shortly afterwards. Sure enough, between 1970 and 1990 the world used 600 billion barrels of oil. So, according to the Club of Rome, reserves should have been overdrawn by 50 billion barrels by 1990. In fact, by 1990 unexploited reserves amounted to 900 billion barrels—not counting the tar shales, of which a single deposit in Alberta contains more than 550 billion barrels.

The Club of Rome made similarly wrong predictions about natural gas, silver, tin, uranium, aluminum, copper, lead and zinc. In every case, it said finite reserves of these minerals were approaching exhaustion and prices would rise steeply. In every case except tin, known reserves have actually grown since the Club's report; in some cases they have quadrupled. "Limits to Growth" simply

Reprinted, by permission, from *The Economist*, "Plenty of Gloom," December 20, 1997. Copyright ©1997, The Economist, Ltd. Distributed by New York Times Special Features/Syndication Sales.

misunderstood the meaning of the word "reserves". . . .

Others have yet to cotton on. The 1983 edition of a British GCSE [General Certificate of Secondary Education] school textbook said zinc reserves would last ten years and natural gas 30 years. By 1993, the author had wisely removed references to zinc (rather than explain why it had not run out), and he gave natural gas 50 years, which mocked his forecast of ten years earlier. But still not a word about price, the misleading nature of quoted "reserves" or substitutability.

## Fallacies About Food

So much for minerals. The record of mispredicted food supplies is even worse. Consider two quotations from Paul Ehrlich's best-selling books in the 1970s.

"Agricultural experts state that a tripling of the food supply of the world will be necessary in the next 30 years or so, if the 6 or 7 billion people who may be alive in the year 2000 are to be adequately fed. Theoretically such an increase might be possible, but it is becoming increasingly clear that it is totally impossible in practice."

"The battle to feed humanity is over. In the 1970s the world will undergo famines—hundreds of millions of people are going to starve to death."

He was not alone. Lester Brown of the Worldwatch Institute began predicting in 1973 that population would soon outstrip food production, and he still does so every time there is a temporary increase in wheat prices. In 1994, after 21 years of being wrong, he said: "After 40 years of record food production gains, output per person has reversed with unanticipated abruptness." Two bumper harvests followed and the price of wheat fell to record lows. Yet Mr Brown's pessimism remains as impregnable to facts as his views are popular with newspapers.

The facts on world food production are truly startling for those who have heard only the doomsayers' views. Since 1961, the population of the world has almost doubled, but food production has more than doubled. As a result, food production per head has risen by 20% since 1961. Nor is this improvement confined to rich countries. According to the Food and Agriculture Organisation, calories consumed per capita per day are 27% higher in the third world than they were in 1963. Deaths from famine, starvation and malnutrition are fewer than ever before.

*"For reasons of their own, pressure groups, journalists and fame-seekers will no doubt continue to peddle ecological catastrophes at an undiminishing speed."*

"Global 2000" was a report to the president of the United States written in 1980 by a committee of the great and the good. It was so influential that it caused one CNN producer to "switch from being an objective journalist to an advocate" of environmental doom. "Global 2000" predicted that population would increase faster than world food production, so that food prices would rise by between 35% and 115% by 2000. As of 1997, the world food commodity index had fallen by 50%. . . .

Perhaps the reader thinks the tone of this viewpoint a little unforgiving. These predictions may have been spectacularly wrong, but they were well-meant. But in that case, those quoted would readily admit their error, which they do not. It was not impossible to be right at the time. There were people who in 1970 predicted abundant food, who in 1975 predicted cheap oil, who in 1980 predicted cheaper and more abundant minerals. Today those people—among them Norman Macrae of the *Economist*, Julian Simon, Aaron Wildavsky—are ignored by the press and vilified by the environmental movement. For being right, they are called "right-wing". The truth can be a bitter medicine to swallow. . . .

## The Latest Environmental Exaggerations

Today the mother of all environmental scares is global warming. Here the jury is still out, though not according to President Bill Clinton. But before you rush to join the consensus he has declared, compare two quotations. The first comes from *Newsweek* in 1975: "Meteorologists disagree about the cause and extent of the cooling trend . . . But they are almost unanimous in the view that the trend will reduce agricultural productivity for the rest of the century." The second comes from Vice-President Al Gore in 1992: "Scientists concluded—almost unanimously—that global warming is real and the time to act is now."

There are ample other causes for alarmism for the dedicated pessimist as the twentieth century's end nears. The extinction of elephants, the threat of mad-cow disease, outbreaks of the Ebola virus, and chemicals that mimic sex hormones are all fashion-

*"Just one environmental scare in the past 30 years bears out the most alarmist predictions made at the time."*

able. These come in a different category from the scares cited above. The trend in each is undoubtedly not benign, but it is exaggerated.

In 1984 the United Nations asserted that the desert was swallowing 21m hectares of land every year. That claim has been comprehensively demolished. There has been and is no net advance of the desert at all. In 1992 Mr Gore asserted that 20% of the Amazon had been deforested and that deforestation continued at the rate of 80m hectares a year. The true figures are now agreed to be 9% and 21m hectares a year gross at its peak in the 1980s, falling to about 10m hectares a year now.

Just one environmental scare in the past 30 years bears out the most alarmist predictions made at the time: the effect of DDT (a pesticide) on birds of prey, otters and some other predatory animals. Every other environmental scare has been either wrong or badly exaggerated. Will you believe the next one?

Environmental scare stories now follow such a predictable line that we can chart their course. Year 1 is the year of the scientist, who discovers some potential threat. Year 2 is the year of the journalist, who oversimplifies and exaggerates it. Only now, in year 3, do the environmentalists join the bandwagon (almost no green scare has been started by greens). They polarise the issue. Either you agree that the world is about to come to an end and are fired by righteous indignation, or you are a paid lackey of big business.

Year 4 is the year of the bureaucrat. A conference is mooted, keeping public officials well supplied with club-class tickets and limelight. This diverts the argument from science to regulation. A totemic "target" is the key feature: 30% reductions in sulphur emissions; stabilisation of greenhouse gases at 1990 levels; 140,000 ritually slaughtered healthy British cows.

*"You can be in favour of the environment without being a pessimist."*

Year 5 is the time to pick a villain and gang up on him. It is usually America (global warming) or Britain (acid rain), but Russia (CFCs and ozone) or Brazil (deforestation) have had their day. Year 6 is the time for the sceptic who says the scare is exaggerated. This drives greens into paroxysms of pious rage. "How dare you give space to fringe views?" cry these once-fringe people to newspaper editors. But by now the scientist who first gave the warning is often embarrassingly to be found among the sceptics. Roger Revelle, nickname "Dr Greenhouse", who fired Al Gore with global warming evangelism, wrote just before his death in 1991: "The scientific basis for greenhouse warming is too uncertain to justify drastic action at this time."

Year 7 is the year of the quiet climbdown. Without fanfare, the official consensus estimate of the size of the problem is shrunk. Thus, when nobody was looking, the population "explosion" became an asymptotic rise to a maximum of just 15 billion; this was then downgraded to 12 billion, then less than 10 billion. That means population will never double again. Greenhouse warming was originally going to be "uncontrolled". Then it was going to be 2.5–4 degrees in a century. Then it became 1.5–3 degrees (according to the United Nations). In two years, elephants went from imminent danger of extinction to badly in need of contraception (the facts did not change, the reporting did).

## Exaggerations Can Be Dangerous

Is it not a good thing to exaggerate the potential ecological problems the world faces rather than underplay them? Not necessarily. A new book edited by Melissa Leach and Robin Mearns at the University of Sussex (*The Lie of the Land*, published by James Currey/Heinemann) documents just how damaging the myth of deforestation and population pressure has been in parts of the Sahel. Westerners have forced inappropriate measures on puzzled local inhabitants in order to meet activists' preconceived notions of environmental change. The myth that oil and gas will imminently run out, together with worries about the greenhouse effect, is responsible for the despoliation of wild landscapes in

Wales and Denmark by ugly, subsidised and therefore ultimately job-destroying wind farms. School textbooks are counsels of despair and guilt (see *Environmental Education*, published by the Institute of Economic Affairs), which offer no hope of winning the war against famine, disease and pollution, thereby inducing fatalism rather than determination.

Above all, the exaggeration of the population explosion leads to a form of misanthropy that comes dangerously close to fascism. The aforementioned Dr Ehrlich is an unashamed believer in the need for coerced family planning. His fellow eco-guru, Garrett Hardin, has said that "freedom to breed is intolerable". If you think population is "out of control" you might be tempted to agree to such drastic curtailments of liberty. But if you know that the graph is flattening, you might take a more tolerant view of your fellow human beings.

You can be in favour of the environment without being a pessimist. There ought to be room in the environmental movement for those who think that technology and economic freedom will make the world cleaner and will also take the pressure off endangered species. But at the moment such optimists are distinctly unwelcome among environmentalists. Dr Ehrlich likes to call economic growth the creed of the cancer cell. He is not alone. Sir Crispin Tickell calls economics "not so much dismal as half-witted".

Environmentalists are quick to accuse their opponents in business of having vested interests. But their own incomes, their advancement, their fame and their very existence can depend on supporting the most alarming versions of every environmental scare. "The whole aim of practical politics", said H.L. Mencken, "is to keep the populace alarmed—and hence clamorous to be led to safety—by menacing it with an endless series of hobgoblins, all of them imaginary." Mencken's forecast, at least, appears to have been correct.

# The Threat of Global Warming Has Been Exaggerated

**by Robert C. Balling Jr.**

**About the author:** *Robert C. Balling Jr. is the director of the office of climatology at Arizona State University in Tempe, Arizona. He is the author of several books on climate change.*

I believe that the best available evidence argues strongly against any rapid and substantial changes to the planetary temperature. Since 1989, a fascinating spectrum of opinions has emerged in the global warming debate. On one end of this spectrum are scientists and some policymakers suggesting that an increase in greenhouse gases will not create any *catastrophic* climate changes in the decades to come. Their assessment leads to the conclusion that the most probable climatic changes (for example, increasing nighttime temperatures, lowering afternoon temperatures, increasing precipitation) may not be disastrous and could even be beneficial to life on the planet.

## A Spectrum of Opinions

Some scientists at this end of the greenhouse-opinion spectrum note that carbon dioxide ($CO_2$) is not a pollutant at all but rather a valuable fertilizer for the growth of plants. Their view is that increasing concentrations of atmospheric $CO_2$ levels may result in the direct benefit of increasing productivity throughout much of the biosphere. The scientists who are at this end of the spectrum tend to be driven by data-based arguments—they seem to be more impressed with the facts than with the predictions from theoretical models.

From a policy perspective, many of these same scientists feel that no "corrective" policy is needed at this time; there is no urgency to rush into immediate policy action and that any realistic policies are likely to fail in an attempt to substantially reduce greenhouse gas emissions. In Vice President Al Gore's

Robert C. Balling Jr., "Global Warming: The Gore Vision Versus Climate Reality," in *Environmental Gore: A Constructive Response to Earth in the Balance*, edited by John A. Baden (San Francisco: Pacific Research Institute for Public Policy, 1994). Reprinted with permission of Pacific Research Institute for Public Policy.

*Earth in the Balance,* we are led repeatedly to believe that these greenhouse skeptics are: (a) small in number and shrinking in number, (b) less credible than other scientists working on the issue, (c) puppets of industry, (d) over-exposed by the media, and (e) looking for "excuses for procrastination."

At the other end of the greenhouse spectrum we find scientists, many policymakers, and much of the environmental community (including assorted movie stars and rock stars) who see a very real disaster in the

> *"I believe that the best available evidence argues strongly against any rapid and substantial changes to the planetary temperature."*

winds. This group believes that increasing $CO_2$ levels will increase planetary temperatures considerably, melting ice caps and raising sea levels around the world. Future droughts will more frequently ravage many agricultural heartlands, super-sized hurricanes will tear away at our coastlines, climate as we know it will change its present day regional structure, and general social and economic chaos will result. Any biological benefits of increasing $CO_2$ will be lost as the biosphere struggles to cope with the rapidly unfolding changes in regional climate. The individuals at this end of the greenhouse-opinion spectrum feel that corrective policy is needed immediately, and they are confident that policy can work to significantly slow greenhouse gas emissions. Quite clearly, the many themes in Gore's *Earth in the Balance* are driven by the underlying beliefs of those who see the greenhouse effect as a severe and immediate threat to the planet; his book is very well received by those at this end of the spectrum.

## Two Views on the Climate System

As a climatologist, I feel confident that an entire volume could be written addressing the many important climate-related points raised throughout Gore's book. To that end, I would strongly encourage readers to examine my own book *The Heated Debate* before or after reading *Earth in the Balance.* You will find that one major, fundamental difference exists between Gore and me—Gore clearly sees the climate system as a collection of fragile but highly interrelated components. I see the climate system as a collection of robust and highly integrated components.

Over billions of years of earth-atmosphere evolution, fragile systems surely would have been replaced by more robust ones; a system of interdependent and fragile components would have little chance of surviving the eons. Gore believes that human-induced changes to atmospheric chemistry are likely, if not certain, to lead to significant changes in climate comparable to those observed over geological time scales. I firmly believe that the earth-atmosphere system will be able to cope with the human-induced changes (which are actually quite small compared to changes over geological time scales) without, as Gore says, throwing the climate system "out of whack."

## The Truth About the Greenhouse Effect

Gore raises two central themes about climate change. First, Gore believes that global climate changes are likely to be substantial and rapid as we continue to add greenhouse gases to the atmosphere. Second, he believes that corrective policies are likely to have a substantial impact on future atmospheric $CO_2$ levels and resulting global temperatures. Here again, I believe that the best evidence available does not support Gore's contention. . . .

No one denies that human activities are increasing the atmospheric concentration of various greenhouse gases. We know with great confidence that due largely to fossil fuel burning and tropical deforestation atmospheric $CO_2$ levels have increased from approximately 295 parts per million (ppm) in 1900 to near 360 ppm in 1994. These molecules of $CO_2$ have the ability to absorb some infrared radiation (energy from the heat of the earth and atmosphere) that otherwise would escape to space. This trapping of heat energy has been likened to the effects found in glass-enclosed greenhouses; hence, $CO_2$ is referred to as a greenhouse gas. However, other gases are also increasing in atmospheric concentration due to human activities, and like $CO_2$, some of these gases are acting to trap additional heat energy. Most important of these other greenhouse gases are methane, various chlorofluorocarbons, and nitrous oxide. When the thermal effects of these other greenhouse gases are expressed in carbon dioxide equivalents, we find that equivalent $CO_2$ has risen from approximately 310 ppm in 1900 to over 430 ppm in 1994.

> *"Given the observed changes in global atmospheric chemistry over the past century, many skeptics ask, 'Where is all the warming?'"*

Herein lies a critical point in the greenhouse debate, and a central reason why the "skeptics" remain important players in the global warming controversy. Given the observed changes in global atmospheric chemistry over the past century, many skeptics ask, "Where is all the warming?" Gigantic numerical models of climate that generate Gore's predicted outcome for continued buildup of greenhouse gases can be used to simulate what should have been observed over the past century. Without fail, these large computer global climate models predicting future disaster suggest that the 40 percent rise in equivalent $CO_2$ seen over the past century already should have produced a global temperature rise somewhere between 1°C and 2°C. Gore claims that global atmospheric temperature has increased by almost 1°C over the past 100 years and that the "upward trend appears to be accelerating as $CO_2$ concentrations increase." If Gore is correct, the global warming theory and the observations would be in general agreement, and many of the skeptics would probably convert to his position immediately. But if Gore is wrong on this important point, a substantial underpinning of his massive Global Marshall Plan would be seriously eroded. It is on this central issue that I strongly believe that Gore is simply dead wrong.

## Lessons from Planetary Temperature

Determining a time series of planetary temperature is no simple task, but fortunately scientists at the University of East Anglia have produced such a time series that has been widely used by climate scientists on all sides of the greenhouse debate. Jones and his colleagues searched millions of temperature records from land-based and ocean-based stations (including ship reports). They assembled the station-specific time series of monthly temperatures, converted all station data to anomalies (deviations from normal monthly temperatures), and interpolated the temperature data onto a 5° latitude by 10° longitude grid system. These gridded temperature anomalies may then be averaged using a weighting system that accounts for the area of the earth represented by each grid point. . . . With regard to this important depiction of global temperature patterns, the following five points argue very strongly against the greenhouse scare.

First, over the period 1890–1990, the linear rise in planetary temperature was 0.45°C, not the "almost" 1°C claimed by Gore. Even if all of the 0.45°C was forced by the buildup of greenhouse gases, the observational record would be pointing to a much smaller greenhouse effect than the one described by Gore. Climate models predicting a greenhouse catastrophe inevitably show that a warming of at least 1°C already should be evident in the temperature record of the past century, and yet the best data we have show less that 0.5°C over this time period.

Second, climatologists are fully aware that urbanization taking place around the globe has created a contaminant to the temperature record. As cities grow, they inevitably get warmer, and thermometers in the urban environment will display a rise in temperature through time that could be mistaken for a greenhouse signal. Somewhere between 5 percent and 25 percent of the global warming of the past century is not real at all—it is a trend that comes from the urban data that contaminates the temperature record.

In addition, overgrazing and desertification, which are taking place over substantial parts of the land area of the planet, have been shown to produce a warming effect that could be mistaken as a greenhouse signal. Overgrazing reduces sparse vegetation in arid and semiarid areas, exposing the thin soils to rapid erosion. Rain that falls in overgrazed areas more quickly runs off the surface, and the incoming sunlight then heats the ground and lower parts of the atmosphere at an accelerated rate. Areas where desertification and overgrazing do not take place will retain the vegetation and soil cover. Here, precipitation does not run off so quickly, and solar energy is used in evaporation and transpiration processes. Less energy goes to heat the ground and air, and near surface air temperatures remain lower than in areas

> *"There is certainly no guarantee that all (or any) of the warming has been forced by the known buildup of greenhouse gases."*

that have been overgrazed. Consequently, areas undergoing overgrazing and desertification are likely to warm through time, and the warming has nothing to do with the buildup of greenhouse gases. Rather, it is due to human-induced activities. Therefore, while the thermometers of the world may show some warming, there is certainly no guarantee that all (or any) of the warming has been forced by the known buildup of greenhouse gases.

## Volcanic and Solar Influences

Third, planetary temperature is driven by a multitude of factors other than the amount of greenhouse gas in the atmosphere. In particular, solar output and volcanic eruptions are two widely discussed candidates for controlling variations in global temperatures.

Most scientists have agreed that large volcanic eruptions eject massive amounts of dust into the stratosphere that can block incoming radiation. When this occurs, less solar radiation reaches the ground, and the planetary temperature drops significantly within a few months of the large eruption. Basic climatological theory predicts cooling from these eruptions, and indeed, the observational global temperature record is in general agreement with the theoretical predictions. Throughout Gore's chapter three, he often refers to this strong linkage between global temperature and volcanic eruptions as evidence of the sensitivity of climate to changes in atmospheric constituents. However, if we statistically control for known variations in stratospheric dust from volcanic eruptions, about 25 percent of the trend in planetary temperature is accounted for over the past century. While I totally agree with Gore's assessment that volcanoes control regional, hemispheric, and global climate, I must add that the volcanoes have been responsible, in part, for the observed trends in planetary temperature over the past century.

> *"Urbanization, overgrazing/ desertification, volcanic dust in the stratosphere, and variations in solar output have influenced planetary temperature over the past century."*

In addition, varying output from the sun should have an obvious impact on the temperature of the earth. However, in the eyes of many scientists, the magnitude of solar output variations does not appear to be sufficient to have caused the observed variations in planetary temperature. Indeed, in referring to scientists working on the solar-climate linkage, Gore states that "neither measurements of the sun's radiation nor the accepted understanding of solar physics lends any credence whatsoever to their speculation." In my own discussion of this topic in *The Heated Debate*, I also concluded that the correlation between solar output and temperature has not been particularly strong at the time scale of a century. However, recent results by Eigil Friis-Christensen and Knud Lassen show a strong correlation between the earth's temperature and the length of the solar sunspot cycle (yet another measure of solar intensity). While no one claims to fully understand the physical connection behind the statistical correlation, this remarkable finding has invigorated the debate on the role of solar variations on earth temperature. Irrespective of the outcome of this debate, it should be apparent that urbanization, overgrazing/desertification, volcanic dust in the stratosphere, and variations in solar output have influenced planetary temperature over the past century. Any claim by Gore or others that the warming trend of the past century is uniquely related to the buildup of greenhouse gases dismisses this compelling evidence to the contrary.

## The Myth of Recent Warming

Fourth, Gore states that the planetary temperature has shown an accelerated increase in the rate of warming in recent years that has coincided with the increase in the atmospheric concentration of greenhouse gases. However, rather simple analysis of the planetary temperatures over the past century shows that between 1890 and 1940 the rise in global temperature was 0.34°C—fully 75 percent of the total warming of the past century (0.45°C) occurred in the first half of the period! A considerable mismatch clearly exists between the observed timing of the warming and the observed buildup in greenhouse gases.

And fifth, a new measurement of the temperature of the earth further argues against Gore's claim of rapid warming in recent years. A fleet of polar-orbiting satellites carries equipment capable of measuring thermal emission of molecular oxygen in the lower part of the atmosphere. This emission of energy is heavily dependent on the temperature of the lower atmosphere, and the emitted energy can escape into space with little disruption from the state of the atmosphere. The satellite-based temperature measurement provides total global coverage, it avoids urban-biases, and appears to be the most accurate method available at this time for computing the true global temperature.

From 1979 to 1990, and during a time of most rapid buildup in atmospheric concentrations of greenhouse gases, the satellite-based temperature measurements have shown a planetary warming of only 0.001°C (inclusion of the most recent two years of record would lower this value due to the cooling effects of the eruption of Mount Pinatubo). Many of the numerical models of climate suggest that the warming (given the known increase in equivalent $CO_2$) should be on the order of 0.3°C over that same time period, and Gore's figure shows warming of approximately that amount. Unfortunately for the pro-greenhouse argument, the satellites are seeing virtually no warming at all, and certainly they are not supporting the claim of accelerated warming in recent decades.

From the evidence above, along with mounds of hemispheric and regional evidence not covered here, I firmly believe that the observed changes in planetary temperature are *not* broadly consistent with expected changes given the known increases in the atmospheric concentration of various greenhouse gases. Most of the observed warming occurred before the bulk of the greenhouse gases were added to the atmosphere, the amount of warming has been too low to be consistent with Gore's catastrophic predictions, and many factors other than the $CO_2$ rise account for the trend and variations in planetary temperature. In addition, this warming has not occurred in the right places (for example, the Arctic region) to be consistent with the models, and as discussed by Patrick J. Michaels, most of the warming has occurred at night (not a greenhouse expectation). Very simply, the climate record over the last century or decade is not pointing in the direction of a greenhouse apocalypse.

# Environmental Scarcity Is a Myth

**by Stephen Budiansky**

**About the author:** *Stephen Budiansky is a writer for* U.S. News & World Report *and the author of* Nature's Keepers: The New Science of Nature Management.

It has become a virtual article of faith that the Earth's population is about to surpass the planet's "carrying capacity." Ecological collapse looms; the only hope is an aggressive effort to reduce runaway birthrates. Lester Brown, president of the Worldwatch Institute, says the "day of reckoning" has already arrived as soil erodes, aquifers empty, pesticide pollution spreads and range lands are overgrazed. "I personally do not think we are ever going to get close" to a world population of 10 billion, Brown told the Senate Appropriations Committee in 1994. The reason? "Ecosystems are already starting to break down," he says.

## Apocalyptic Visions

Even President Bill Clinton has joined the neo-Malthusian bandwagon; he was riveted by an apocalyptic jeremiad that appeared in February 1994 in the *Atlantic Monthly*. The piece, written by foreign correspondent Robert Kaplan, envisions a world of growing chaos, anarchy, disease and corruption as hungry refugees surge across borders in search of food and nations fight over scarce resources. Humanitarian disasters such as the one in Rwanda are a herald of the new era of resource limits.

But if these apocalyptic prophecies come true, it will not be simply because man has been too fruitful and has been multiplying too fast. True, Rwanda was the most densely populated country in Africa before the civil war erupted. But its Hutu and Tutsi peoples are battling over tribal hatreds and political power, not resources: Rwanda was about to reap a copious harvest when the killing started.

Recent scientific studies confirm that the Earth's basic resources are vastly greater than what are needed to feed even the 10 billion people who are almost certain to inhabit the planet by the middle of the twenty-first century. The real threat is not that the Earth will run out of land, topsoil or water but that nations

Reprinted, by permission, from Stephen Budiansky, "Ten Billion for Dinner, Please," *U.S. News & World Report*, September 12, 1994. Copyright ©1994, U.S. News & World Report.

will fail to pursue the economic, trade and research policies that *can* increase the production of food, limit environmental damage and ensure that resources reach the people who need them. Indeed, embracing the myth of environmental scarcity could ironically prompt the United States and other countries to adopt policies that virtually guarantee that the apocalyptic future that environmentalists foretell really does come true.

## Population Control and Food Supply

The conventional solutions to the world food problem, heavily publicized . . . by groups such as Zero Population Growth and the Pew Global Stewardship Initiative, . . . focus almost exclusively on population control: They argue that further major increases in food production are not possible. "Achieving a humane balance between food and people now depends more on family planners than on farmers," says Worldwatch's Brown, who cites an apparent drop-off in per capita grain production as proof that the era of rapid technological progress "has slowed to a trickle."

The U.S. government has also been heading in this direction. The Agency for International Development has been cutting funding for agricultural research for a decade; in 1993 its contributions to international agricultural research programs fell by almost one third. . . .

> *"The real threat is . . . that nations will fail to pursue the economic, trade and research policies that* can *increase the production of food."*

Reducing birth rates is clearly an urgent priority if the world population is to have even a chance of stabilizing at twice current numbers. But even the most draconian birth control efforts cannot prevent Earth's population from almost doubling. And that creates an unprecedented challenge that can be met only by increasing food production.

## Food Production Has Grown

Despite the doomsayers, agricultural production is still on a rapid upward trend. Most telling is the fact that the world price of food has been declining for the past 50 years. Since 1970, it has dropped by one half in real terms, hardly what one would expect of a dwindling resource.

The growth in world grain production has slowed since 1990. But by falsely attributing that slowdown to technological or resource limits, environmentalists have greatly exaggerated the obstacles to future yield improvements. Grain is in surplus, and 46 million acres of U.S. farmland and 11 million acres in Europe have been deliberately idled under government programs to boost farmers' incomes. Close to 200 million acres of South American savanna also could be brought into production if demand, and free trade, permitted.

Total food production, meanwhile, continues to grow much faster than population. Per capita world food output grew by 5 percent during the 1980s. "The world has not reached, nor is it near, the upper limits of production capacity," says B.H. Robinson of the U.S. Department of Agriculture's Economic Research Service. With sufficient investments in agricultural research and the removal of trade barriers, agricultural economist Dennis Avery of the Hudson Institute told a Senate committee in 1994, the Earth should readily be able to feed 10 billion people.

> *"Total food production . . . continues to grow much faster than population."*

A study by Paul Waggoner of the Connecticut Agricultural Experiment Station in New Haven backs up the claim that food production is for the foreseeable future limited only by human ingenuity, not natural resources. The gross productive potential of the Earth—set by available land, climate and sunlight for photosynthesis—is sufficient to produce food for a staggering 1,000 billion people. Even without irrigation, available water is sufficient to grow food for 400 billion, and a conservative estimate of sustainable fertilizer production implies ample supplies to produce food for 80 billion.

## The Need for Agricultural Technology

The technological potential for further yield gains is demonstrated most dramatically in the case of corn. Without any contribution from genetic engineering or other futuristic technologies, corn yields per acre are increasing 1 to 2 percent a year—despite the fact that yields have already increased by a factor of 10 since intensive breeding of corn began early in this century. "If anything should have played out, corn should have," says Ralph Hardy, president of Cornell University's Boyce Thompson Institute, a leading center of plant biotechnology research. The highest corn yields achieved each year in Iowa are nearly five times the world average—and they are increasing at a rate of 125 pounds per acre each year, more than double the average world gain.

Whether worldwide yields can continue to increase at the same rate—as they must if 10 billion people are to be fed—depends not on the ecological carrying capacity of Spaceship Earth, agricultural researchers argue, but on something more prosaic: money to fund continuing research and development of new technologies, and credit and technical assistance so that farmers in the developing world can apply the technologies that already exist. It would be impossible to feed the world's current population with the technology of the 1960s—at least without a devastating impact on the environment. And for three decades, the remarkable successes of the Green Revolution have come principally from a worldwide network of agricultural research centers that operate under the Consultative Group on International Agricultural Research (CGIAR).

The International Rice Research Institute in the Philippines, for example, developed genetically improved strains of rice that have doubled yields per acre.

New varieties now under development concentrate an extra 10 percent of their energy into developing their grain-bearing seed heads and add characteristics such as drought, flood, salt and pest resistance that can further boost yields.

"The *only* reason we've had these large yield increases is that we've made large investments in the past," says Per Pinstrup-Anderson, director of the International Food Policy Research Institute. "It's been fine up until now only because people with foresight did the right thing." But funding for the CGIAR centers has begun falling. A recent infusion of cash from the World Bank to make up a $55 million shortfall in their research budget has helped, but the centers have already lost 100 scientists since 1992. U.S. aid for foreign agricultural programs has been on a steep decline, falling by three quarters since 1980.

## Do Not Blame the Western World

If the world is to feed its doubling population, nations also must remove agricultural trade barriers—another policy that many environmental groups and developing nations are working to thwart. In the conventional wisdom, curbing the "excess consumption" of Western countries goes hand in hand with population control; many environmental activists link high-yield agriculture, affluence, free trade and technological advancement with environmental devastation. A Worldwatch Institute report, for example, complains that the benefits of world trade in spreading the wealth have been "overwhelmed" by its tendency to encourage "unsustainable consumption by creating the illusion of infinite supplies." Under Secretary of State for Global Affairs Timothy Wirth . . . has echoed this theme, promising "to take on the difficult issue of wasteful resource consumption and the disproportionate impact the developed world has on the Earth's environment."

Reducing the environmental strains caused by the profligate use of resources—wherever it occurs—is only common sense. But given the world surplus of grain and falling food prices, many experts say it is ludicrous to blame the hunger that plagues parts of the world on greedy Western lifestyles. "This is the Congregational Church syndrome," says Robert Repetto of the World Resources Institute. "People think if they bring some canned goods to church on Sunday it will relieve world hunger. It doesn't work that way."

Malnutrition afflicts 700 million people not because the world does not produce enough food but because poor nations do not have enough money to buy it. Nor is the poverty of the Third World simply a reflection of the rapaciousness of the First World. Despite the often-quoted statistics about how the United States consumes one quarter of the world's energy and produces a similarly disproportionate share of carbon dioxide and other pollutants, the truth is that the United States and other advanced countries

> *"The most environmentally sound way to increase world food production rapidly is to expand it in places where the soils can sustain it."*

use far less energy and produce far less carbon dioxide per dollar of GDP added to the world's economy. They also have, for the most part, far more stringent environmental policies than many developing nations.

## Unsound Policies

In India, for example, the government provided farmers with free electricity, which encouraged them to overpump water, leading to salinization and waterlogging of soils. The clearing of Brazilian rain forests and the farming of these fragile soils have likewise been heavily encouraged by government subsidies. In fact, many of the most publicized crises in agricultural production—salt-clogged soils, overpumped aquifers, eroding soils—reflect not fundamental resource limits but bad resource management. "The most important element driving the conversion [of forest to farming] is lack of jobs," says Repetto—not lack of food.

Far from encouraging environmental devastation, freer trade in agriculture could actually help ensure that increased food production is concentrated on what Dennis Avery of the Hudson Institute calls "the best and safest acres." Soils in tropical climates are extremely poor in inherent fertility; the most environmentally sound way to increase world food production rapidly is to expand it in places where the soils can sustain it and where high-yield technologies and good transportation are already available—which means the world's temperate zones.

A free market would discourage inefficient food production on farmland hacked out of tropical rain forests. Yet policies of "self-sufficiency" and protectionism encourage just such environmentally unsound and unsustainable policies. Indonesia, for example, is clearing 1.5 million acres of tropical forest to grow soybeans for chicken feed at a cost above the world price; India produces milk at two to three times the world price. On the other hand, many nations that have made the transition from agricultural to industrial economies are far from "self-sufficient"; Japan, for example, imports three-fourths of its grain.

But most of all, says Pinstrup-Anderson of the International Food Policy Research Institute: "It's not intensive agriculture that's causing environmental degradation in the developing countries, it's a lack of it." Advanced intensive methods of cultivation have dramatically reduced the amount of land, water, soil and energy required to produce a ton of grain—breaking the link between economic growth and environmental impact. While poverty and a lack of intensive production methods have forced millions of Third World farmers to overgraze marginal range land or plow up steep hillsides with primitive methods, modern techniques such as "no-till" farming that have been widely adopted in developed countries have cut soil erosion rates dramatically—often virtually to zero—while boosting yields significantly.

"I'm an optimist, no question about it," says Pinstrup-Anderson. But his optimism is tempered by the knowledge that solutions to the world's food problem will not simply drop into our laps—especially given current policy directions. "Complacency is our enemy," he says. So is fatalism.

# The Media Exaggerate Environmental Crises

## by Alan Caruba

**About the author:** *Alan Caruba is a business and science writer. He is the founder of the National Anxiety Center, which monitors the media for scare campaigns.*

Apocalyptic predictions are always a sure bet to capture the public imagination. And because seizing the public's attention is one of the media's driving goals, such predictions are often in the news.

This is certainly true of the environmentalist movement. Since its advent in the 1970s, the movement has generated countless headlines about global warming, disappearing ozone and forests, and dire health threats posed by asbestos, radon, and even water.

### The Manipulation of the Media

In the United States and around the world, people are highly dependent on the media but are also particularly vulnerable when the information is inaccurate. Few editors and reporters have a background in science with which to make informed decisions, and thus they, too, are easily manipulated.

Since the first Earth Day in 1970, a vast public relations campaign has been in place to influence public support and generate environmentally friendly legislation—which has had the collateral effect of slowing economic growth.

Fully 30 percent of all federal regulations have been generated by the Environmental Protection Agency. When the EPA was established in 1970, it had a budget of $205 million and a staff of about 4,500. Today, it controls a budget of $4.3 billion, and its staff has quadrupled to almost 18,000.

### The Price of Environmental Regulations

According to EPA estimates, the cost to citizens and the states of responding to regulatory mandates is about $100 billion a year. Independent analysts have put the figure at more than $500 billion annually.

Reprinted, by permission, from Alan Caruba, "The Many Excesses of Ecojournalism," *The World & I*, July 1995, by permission of the author.

For example, in 1991, Columbus, Ohio, commissioned a study of federal and state environmental requirements as they impinge on that city's finances. The study found that, in 1995, nearly 20 percent of Columbus' total budget would have to be spent just to comply with existing regulations and other unfunded mandates. That adds up to more than $100 million in just one city.

Critics of regulation point out how many additional police officers or fire fighters could be hired for that amount, how many roads and bridges could be resurfaced and upgraded, and how many computers could be purchased for schools.

## Apocalyptic Falsehoods

In general, the environmentalist media campaign has been so successful it seems that today's news is often determined to convince people that much of what they eat, drink, and breathe is life threatening. Other apocalyptic stories focus on global warming, the loss of the protective ozone layer, a worldwide population explosion, and even the occasional rogue asteroid.

There is, however, an inherent problem with catastrophic claims and predictions that warn of dire threats to the public health. For example, in America and around the world, fertility rates have been steadily declining. Thus, the true cause for population growth is the fact that people are living longer, healthier lives. Indeed, in America life expectancy has reached 75.8 years, the highest it has ever been.

> *"It seems that today's news is often determined to convince people that much of what they eat, drink, and breathe is life threatening."*

Right now, in schools across America, curriculums are larded with environmental doomsday scenarios. The EPA's Office of Environmental Education has a $65 million budget. It is only one of many organizations striving to convince young Americans of the one-sided view that the earth is (a) fragile and (b) doomed because of pollution by human beings.

## Ozone Layer Hysteria

A sharply differing view of the earth and pollution, held by many reputable scientists, is poorly reported by the media. Let's look, for example, at the claim that the upper atmosphere's ozone layer, which protects against the cancer-causing effects of ultraviolet radiation, is severely threatened.

Environmentalists and the journalists who report their claims have convinced the government that all the ozone surrounding the earth is threatened by "holes" caused by chlorine and fluorine from the breakdown of the Freon used in traditional refrigerators and air conditioners. These "holes," or thinner areas of ozone concentration, were discovered back in the 1950s, well before the widespread use of air-conditioning. Thus, many scientists believe that the holes are a completely natural phenomenon.

Nevertheless, these claims have secured a ban on the manufacture of chlorofluorocarbons (CFCs), the chemicals that compose Freon. In part because of journalists' failure to fully investigate and report on ozone hole claims, more than 100 million refrigerators in America are now obsolete, along with 90 million air conditioners in office buildings, homes, and cars.

Environmentalists are currently trying to ban chlorine, a chemical used in the manufacture of a wide variety of commercial products, despite the fact that the cycles of evaporation of the world's oceans put 300 times more chlorine into the atmosphere than the entire world's former production of CFCs. The latter is yet another fact seldom reported in the major media.

## Further Fallacies

Another example of the success of the environmentalist media campaign is the way many people think all the nation's forests are disappearing. The United States, however, is home to two-thirds of the woods that existed when the Pilgrims landed at Plymouth Rock. In U.S. national forests, moreover, annual growth exceeds harvest by 55 percent. And loss of the woods has not been an entirely bad thing. Observers note that one-third of former forests and grasslands are represented by farms on which 2 percent of the population feeds all 250 million Americans.

Another textbook case of the environmentalist spin in the media is the issue of global warming. This widely reported *theory*—not fact—was first advanced in the 1890s and again in the 1950s.

By 1974, however, it was replaced by the theory that a new ice age was beginning. *Time* magazine reported that "the atmosphere has been growing gradually cooler for the past three decades. The trend shows no indication of reversing." This dubious hypothesis about an impending ice age is no longer heard. And concerning the global warming theory, many scientists note that the earth has warmed barely 1 degree Fahrenheit in the past century. If the media were more objective, they would report that warming and cooling cycles are a natural part of the earth's ecosystem and generally require hundreds of years to occur.

Journalists assigned to cover the environment beat are most responsible for advancing these and other environmentally inspired doomsday scenarios. Almost from the beginning, these journalists allied themselves with the environmentalist movement and, for the most part, jettisoned any serious effort to question the various apocalyptic scenarios that were passed along to the public.

> *"Journalists . . . know full well that headlines predicting disasters of one sort or another ensure readership and viewership."*

But why didn't the nation's scientific community speak up more forcefully? Perhaps because scientists like to *prove* things first—and that takes time—and perhaps because scientists are uncomfortable doing public relations.

And why didn't editors exercise more caution with each new environmental scare campaign—about asbestos or radon or endangered species, and so forth? Because many editors know as little as the reporters, or even less.

Ultimately, the success of journalism depends on selling advertising and getting ratings. It is, after all, a market-driven business enterprise. And journalists—reporters, editors, anchorpeople, and producers—know full well that headlines predicting disasters of one sort or another ensure readership and viewership.

> *"The consensus among many scientists . . . is that far too much of what environmental journalists have written is seriously flawed and unbalanced."*

Most journalists are bright people who love a good story. Most, too, have a strong devotion to accuracy and truth. Thus, the most astonishing thing about environmental journalism is the way it was exempt for so long from journalists' ingrained cynicism regarding claims by everyone from the local councilman to the president of the United States.

## The Environmental Agenda

The idealism, activism, and prophetic aura of environmental groups, many of which also quickly became politically powerful, with multimillion-dollar budgets, may have swayed too many journalists in the early years of the movement.

Today, however, environmental groups have seen a significant loss of membership support. Since 1990, Greenpeace has lost 40 percent of its members, the Wilderness Society has suffered a 35 percent decline, and the Sierra Club has slashed its current budget by $4 million and fired 40 members of its staff.

Some journalists have begun to seriously entertain the possibility that environmentalism may be driven by a disdain for the capitalist free-market system that underpins constitutional liberties like property rights. The writings of extremist environmentalists have always lent credence to this possibility.

For example, Michael McCloskey, chairman of the Sierra Club, wrote in 1970 in *Ecotactics:*

> That other revolution, the industrial one, is turning sour, needs to be replaced by a revolution of new attitudes toward growth, goods, space, and living things.

The attitudes he refers to amount to a recognition of the need for a slowing of economic growth.

The introduction to *Ecotactics* was written by consumer activist Ralph Nader. "To deal with a system of oppression and suppression, which characterizes the environmental violence in this country," Nader wrote, "the first priority is to deprive the polluters of their unfounded legitimacy."

Many analysts today say that these views in fact reflect the core beliefs of those who have been setting the environmental agenda for decades.

## Scientific Illiteracy

Such views are aided by the virtual scientific illiteracy of print and electronic media professionals. At a 1992 panel discussion on emerging environmental issues, held during a convention of the Society of Professional Journalism, Kevin Carmody, the metropolitan editor of the Charlottesville, Virginia, *Daily Progress* and secretary of the Society of Environmental Journalists, spoke about a survey of 1,600 newspaper editors, nearly half of whom believed that men and dinosaurs coexisted on earth.

This *Flintstones* understanding of science means that a large number of people responsible for what Americans read have no idea that dinosaurs had died off tens of millions of years before the advent of *Homo sapiens.*

At the third annual convention of the Society of Environmental Journalists in 1993, the group struggled with the realization that there was a growing backlash, both inside newsrooms and among the public, over the accuracy and credibility of their work.

One journalist in attendance said, "Reporters sometimes view their environmental beat as a mission in which the fate of the world and its people is at stake." Others noted a widespread perception of "a left-wing bias" in their work.

A study by the Foundation for American Communications, conducted by American Opinion Research, concluded that only 3 percent of the journalists surveyed considered the overall quality of environmental coverage to be very good. That's 97 percent who rated it fair to poor.

## A Rejection of Apocalyptic Journalism

When Americans confounded a host of political pundits in November 1994 by the decisiveness of their rejection of the status quo [by electing a Republican majority in Congress], they were also voting, many analysts say, against the subversion of property rights, the environmental strangulation of economic growth, and decades of end-of-the-world scenarios.

Many Americans have come to feel they've been ill served by apocalyptic journalism. The letters to the editor pages of American newspapers are filled with protests against the costs and strictures that are in part the result of overzealous, scientifically inaccurate reporting that has initiated laws that continue to affect every business and individual.

The consensus among many scientists—and a growing number of media people—is that far too much of what environmental journalists have written is seriously flawed and unbalanced. So, as the world approaches the millennial year 2000, it is essential that print and broadcast journalists undertake a serious effort to undo the harm that's been done and avoid the pitfalls of apocalyptic claims.

# Chapter 2

# Should Biodiversity Be Preserved?

# Preserving Biodiversity: An Overview

## by Kenneth Jost

**About the author:** *Kenneth Jost covers legal issues for the* CQ Researcher, *a weekly report published by* Congressional Quarterly.

When Congress passed the Endangered Species Act in 1973, it declared that endangered wildlife and plants are of "esthetic, ecological, educational, historical, recreational and scientific value." Supporters of the law continue to stress the value of individual species. But critics complain that the benefits of protecting species are outweighed by the costs in actual spending, economic disruption and political discontent.

### Biodiversity Is Valuable

Biologists write broadly about the benefits of biodiversity in both concrete and less tangible ways. "Biological diversity is the key to the maintenance of the world as we know it," Edward O. Wilson, the celebrated Harvard entomologist, writes in his influential 1993 book *The Diversity of Life.*

In his book, *The Value of Life,* Stephen Kellert of Yale University identifies nine "values" served by biodiversity. Kellert begins with the "utilitarian" value of plants and animals for food, clothing, medicine and more. He then proceeds through less tangible benefits, such as the joy of exploring nature ("naturalistic") and the pleasure of the physical splendor of nature ("aesthetic").

The interest groups that work the endangered species issue in Washington concentrate on the most concrete of these benefits. They have recently emphasized the importance of rare plant species as a source of medicines. The most famous example is the rosy periwinkle of Madagascar, which produces two alkaloids used in manufacturing drugs to treat Hodgkin's disease and acute lymphocytic leukemia.

"Endangered species help save lives," says environmental consultant Randall Snodgrass, a former wildlife director for the National Audubon Society. "So if

Reprinted from Kenneth Jost, "Protecting Endangered Species," *CQ Researcher*, April 19, 1996.

for no other reason than the health of humankind, protecting endangered species is important."

Critics of the law minimize the importance of rare species in new medicines. "Don't expect to find a cure for cancer from some endangered species," says Michael L. Plummer, co-author with science journalist Charles C. Mann of *Noah's Choice,* a critical look at the Endangered Species Act.

> *"Most critics . . . do not challenge the value of biodiversity but rather the priority that [the Endangered Species Act] gives to the goal."*

Still, most critics of the act do not challenge the value of biodiversity but rather the priority that the law gives to the goal. "Grand statements about biodiversity being of utmost importance are true but trivial," says Plummer, who is a fellow at the Discovery Institute, a conservative, Seattle-based environmental think tank. "They don't help you set a strong, balanced policy toward protecting biodiversity."

In actual spending, the cost of the endangered species program is relatively small—about $79.3 million in 1995 for the Fish and Wildlife Service (FWS) and $7.9 million for the smaller, and less controversial program, in the Commerce Department's National Marine Fisheries Service. Snodgrass notes that before the Clinton administration increased the FWS budget, it was spending about $40 million a year—around the cost of building a single mile of interstate highway. "With the millions and millions of dollars we spend on highways," Snodgrass says, "we can afford to do what is necessary to protect the habitat of threatened and endangered species."

But the critics say the act is causing huge economic dislocations, particularly the restrictions on timber harvesting in the Northwest. They say the restrictions, designed to protect the spotted owl, have cost thousands of jobs, higher housing prices and social dislocations in logging communities. "This decimated public timber harvesting in the Northwest," says Robert Nelson, a professor of public policy at the University of Maryland, College Park, and a senior fellow at the Competitive Enterprise Institute.

## Job Loss and the Environment

Some supporters of the law concede that the timber restrictions have hurt employment. But in December 1995, 60 economists in the region said the industry faced a sharp downturn for other reasons, especially the reduction in old-growth forests due to earlier overharvesting.

"The downsizing would have been required regardless of the spotted owl," says Thomas Power, chairman of the economics department at the University of Montana, Bozeman, and principal author of the report. "The role that Endangered Species Act actions solely by themselves have played has been quite modest."

As for broader effects, even critics like Nelson concede that the evidence is slim. "You can't pin the loss of jobs on the Endangered Species Act in other situations," he says. But Plummer still insists that the law has been disruptive. "If the purpose of the law were solely to put a monkey wrench in economic activities that threaten species," he says, "it's done that."

In the most thorough research on the question, however, Stephen Meyer, director of the Massachusetts Institute of Technology's Project on Environmental Studies, found no evidence of harmful economic effects. Meyer's studies compared individual states' economic performance with the number of endangered species in each state and found no effect on overall economic growth, agricultural production or construction activity. "If [the critics'] rhetoric is even partially true, then one should be able to look at the economies of the state and see some effects," Meyer explains. "There are no such impacts."

As for the impact on government projects, the Interior Department says the law has blocked only a very few. Out of 186,000 federal actions reviewed between fiscal 1987 and fiscal 1995, only 600 were found to pose a danger to a species, and only 100 were actually blocked, the department's most recent data show. All but 13 of those involved timber sales in the Pacific Northwest, the department says.

## Responding to the Opposition

Still, critics insist that the law has hurt individual landowners, creating political discontent and undermining support for the legitimate goal of protecting endangered species. "You're creating a class of incredibly pissed off people all across the country," says Smith at the Competitive Enterprise Institute.

Supporters concede the need to deal with the opposition to the law, but they also believe the public will back it when it's properly understood. "We have to try to address effectively the concerns of the populace," says Michael Clegg, chairman of a National Academy of Sciences (NAS) panel that issued a largely favorable assessment of the law in May 1995. "We also have to make sure we're doing a good job of educating the people about the consequences of the loss of biological diversity."

# Biodiversity Should Be Preserved

**by David Langhorst**

**About the author:** *David Langhorst is an executive at the Alabama Wildlife Federation.*

The Endangered Species Act (ESA) has produced a remarkable string of successes. In its history, it has stabilized or improved the conditions of hundreds of plant and animal species that had been in serious decline. In my own work, I have seen large numbers of concerned citizens work with the ESA to help bring about the recovery of the gray wolf in the Northern Rockies ecosystem. By educating communities about the importance of the wolf to the health of the ecosystem and using the ESA's flexible provisions, we are successfully restoring this wonderful animal to the wild in a manner sensitive to local economic interests.

The gray wolf recovery effort is a model of how diverse groups of local citizens can work together to achieve results using the ESA. However, as a result of delaying tactics by narrow ranching interests, wolf recovery is taking too many years and is generating inordinate costs to the Federal taxpayer. Meanwhile, during the period of the wolf recovery effort, the recovery of numerous other listed species is being neglected.

## Two Reasons to Protect the ESA

Certain regulated industry groups are now advocating that the ESA's goal of protecting and recovering all of the Nation's imperiled plant and animal species be abandoned and that the fate of each species be left to the discretion of the Secretaries of Interior and Commerce. Such an abandonment of the ESA's goal would be unwise for at least two reasons. First, conserving the fullest extent of our natural heritage provides enormous benefits to people, benefits that greatly exceed the costs of protection measures. Second, the alternative—separately deciding the fate of each species using a cost/benefit analysis—is simply unnecessary, unworkable, and would be extremely wasteful considering the nu-

Reprinted from David Langhorst, "Is the Endangered Species Act Fundamentally Sound?" *Congressional Digest*, March 1996.

merous ESA procedures already in place to ensure that economic consequences are considered before the law is implemented.

Congress established the goal of protecting and recovering all imperiled species when it first enacted the ESA in 1973. This ambitious goal was not chosen carelessly, but was arrived at after Congress determined that the rapid loss of biodiversity in the U.S. and abroad posed a direct threat to

> *"Conserving the fullest extent of our natural heritage provides enormous benefits to people."*

the well-being of the American people. When the law was reauthorized in 1978, 1982, and 1988, Congress reaffirmed that recovering all threatened and endangered species was essential.

The scientific evidence that motivated previous Congresses to set the goal of recovering all species has only strengthened in recent years. Today there is no dispute in the scientific community that human activity has brought about a loss of biodiversity not witnessed since the cataclysmic changes ending the dinosaur era 65 million years ago. Edmund O. Wilson, the eminent Harvard biologist, estimates that the current extinction rate in the tropical rain forests is somewhere between 1,000 to 10,000 times the rate that would exist without human disturbances of the environment. According to the recent study of the ESA by the National Academy of Sciences, the "current accelerated extinction rate is largely human-caused and is likely to increase rather than decrease in the near future."

This rapid loss of biodiversity is occurring not just in the tropical rain forests. In the nearly 400 years since the Pilgrims arrived to settle in North America, about 500 extinctions of plant and animal species and subspecies have occurred—a rate of extinction already much greater than the natural rate. According to recent calculations by Peter Hoch of the Missouri Botanical Garden, over the next five to 10 years, another 4,000 species in the U.S. alone could become extinct. This evidence of increased extinctions provides sad testimony to the need for improving the ESA rather than scaling back its fundamental goal.

## Benefits to People

Species are essential components of natural life-support systems that provide medicines, food, and other essential materials, regulate local climates and watersheds, and satisfy basic cultural, aesthetic, and spiritual needs. Below are six examples of how endangered species protections help people.

*New Medicines to Respond to the Health Crises of Tomorrow.* Wild plant and animal species are an essential part of the $79 billion annual U.S. pharmaceutical industry. One-fourth of all prescriptions dispensed in the U.S. contain active ingredients extracted from plants. Many other drugs that are now synthesized, such as aspirin, were first discovered in the wild.

Researchers continue to discover new potential applications of wild plants and animals for life-saving or life-enhancing drugs. In fact, many pharmaceutical companies screen wild organisms for their medicinal potential. Yet, to date, less than 10 percent of known plant species have been screened for their medicinal values, and only 1 percent have been intensively investigated. Thus, species protections are essential to ensure that the full panoply of wild plants and animals remains available for study and future use.

*Wild Plant Species that Safeguard our Food Supply.* The human population depends upon only 20 plant species, out of over 80,000 edible plant species, to supply 90 percent of its food. These plants are the product of centuries of genetic cross-breeding among various strains of wild plants. Continual cross-breeding enables these plant species to withstand ever-evolving new diseases, pests, and changes in climatic and soil conditions. According to a study, the constant infusion of genes from wild plant species adds approximately $1 billion per year to U.S. agricultural production.

If abundant wild plant species were unavailable to U.S. agriculture companies for cross-breeding, entire crops would be vulnerable to pests and disease, with potentially devastating repercussions for U.S. farmers, consumers, and the economy.

*Renewable Resources for a Sustainable Future.* At existing levels of consumption, nonrenewable resources such as petroleum will inevitably become increasingly costly and scarce in the coming decades. To prepare the U.S. for the global economy's certain

> *"Our society has become so alienated from nature that sometimes we forget that we rely on ecosystems for our survival."*

transition toward renewable resources, Congress must ensure the health of the U.S. biological resource base. Fish, wildlife, and plant species could potentially supply the ingredients for the products that drive the U.S. economy of the 21st century. The substance that holds mussels to rocks through stormy seas, for example, may hold clues for a better glue to use in applications from shipbuilding to dentistry.

## Ecosystems Are Essential

*Early Warning of Ecosystem Decline.* Scientists have long known that the loss of any one species is a strong warning sign that the ecosystem that supported the species may be in decline. A recent study reported that loss of species could directly curtail the vital services that ecosystems provide to people. A subsequent study suggests that destruction of habitat could lead to the selective extinction of an ecosystem's "best competitors," causing a more substantial loss of ecosystem functions than otherwise would be expected.

Negative impacts in wild species often portend negative impacts for human health and quality of life. For example, some animal species are critical indicators of the harm that heavy chemicals can cause in our environment.

*Ecosystems: Life-Support Systems for People.* Our society has become so alienated from nature that sometimes we forget that we rely on ecosystems for our survival. Ecosystems carry out essentially natural processes, such as those that purify our water and air, create our soil, protect against floods and erosion, and determine our climate.

For example, the Chesapeake Bay, the Nation's largest estuary, not only supports 2,700 plant and animal species, but also plays a major role in regulating environmental quality for humans. Rapid development around the Bay has sent countless tons of sediment downstream, landlocking communities that were once important ports. The construction of seawalls and breakwaters in some areas has led to rapid beach erosion in others.

*Ecosystems: Industries and Jobs Depend on Them.* Healthy ecosystems enable multi-billion dollar, job-intensive industries to survive. Examples of industries that are dependent on the health of ecosystems are: tourism, commercial fishing, recreational fishing, hunting, and wildlife watching.

When ecosystems are degraded, the result is economic distress. Destruction of salmon runs on the Columbia and Snake river systems in the Pacific Northwest led to the near-collapse of that region's multi-billion dollar commercial and sport fishing industries.

## Problems with Cost/Benefit Analysis

Anti-ESA advocates propose to replace the goal of saving all species with a cost/benefit analysis to determine whether to save each species. Such cost/benefit analysis would likely produce an extinction of hundreds of endangered species due to human disturbances of habitat.

In the absence of any legal obligation to recover species, the Secretary of the Interior could ultimately succumb to political pressures and choose meager objectives for any species that dare to get in the way of industry or development. For most species, any objective short of full recovery would effectively perpetuate the continued slide toward extinction.

Even if the cost/benefit analysis could somehow be insulated from political manipulation, its outcome would still be totally unreliable. The information available to the Secretary about the costs of protecting the species in question would be extremely incomplete, because no one could know at the time of the cost/benefit analysis what human activities would ultimately threaten the species and whether those activities could be modified through the ESA consultation process to avoid or reduce economic losses.

*"Despite years of research and development, we have only just begun to discover the beneficial uses of species."*

Equally important, the Secretary would also have incomplete information about the benefits to people provided by the species. Despite years of research and development, we have only just begun to discover the beneficial uses of species. Of the estimated five to 30 million species living today on Earth, scientists have identified and named only about 1.6 million species, and most of these have never been screened for beneficial uses. As species become extinct, we simply don't know what we are losing. The species that become extinct today might have provided the chemical for a miracle cancer treatment or the gene that saves the U.S. wheat crop from the next potentially devastating disease. A cost/benefit analysis of the penicillin fungus in the years prior to the discovery of its antibiotic qualities would have been a surefire recipe for extinction because no one could foresee its future role in the development of wonder drugs that would save and enhance the lives of millions of people.

## The Act Should Be Improved

There is yet another reason why we should not attempt to decide the fate of species based on a prediction of their future benefits. Species within an ecosystem are interdependent, and thus the extinction of one species potentially disrupts other species and the functioning of the entire ecosystem. As reported by the Missouri Botanical Garden, the loss of one plant species can cause a chain reaction leading to the extinction of up to 30 other species, including insects, higher animals, and other plants. Like pulling a single bolt from an airplane wing, we cannot know beforehand what effect the loss of a single species might have on the entire ecosystem.

A final flaw with the cost/benefit approach is that it is based on a false premise that the ESA lacks opportunities for consideration of economic and social impacts of listings. In fact, numerous ESA provisions require that economic and social consequences be balanced with species protection goals. Once a species is listed, the ESA provides for the consideration of socioeconomic factors in the designation of critical habitat, the development of special regulations for threatened species and experimental populations, the issuance of incidental take permits, the development of reasonable and prudent alternatives during Federal agency consultations, and the existence of the Endangered Species Committee to resolve any conflicts between conservation and economic goals.

Reauthorization of an effective Endangered Species Act is in the best interest of everyone involved. Species provide untold benefits to humans and are essential to our quality of life. By making thoughtful improvements to the ESA, we can enable private landowners and other stakeholders to take a greater conservation role and thereby provide for both species conservation and sustainable development—for the benefit of each of us and generations to come.

# Preserving Biodiversity Is a Jewish Obligation

**by Bradley Shavit Artson**

**About the author:** *Bradley Shavit Artson is a rabbi and author.*

> *Of all that the Holy Blessed One created in the world, God created nothing without a purpose.*
> —The Talmud

The wondrous variety of living things is under assault. Precious ecosystems, including 50 percent of the wetlands in the continental United States and 90 percent of the lowland coniferous forests in the Pacific Northwest, are already gone. Species, too, are suffering unprecedented loss: Scientists estimate that from 5 to 20 percent of the tropical forest species will become extinct in the next 30 years. Stuart Pimm notes in the journal *Science* that current extinction rates are 100 to 1,000 times their prehuman levels.

Since everything is dependent on the continuing vitality of the biosphere, this threat to biological diversity not only threatens the human species but also diminishes the vast splendor of creation.

## A Complex Interaction

A meaningful explanation of biodiversity is offered in *The Encyclopedia of the Environment* by Ruth and William Eblen: "Life reveals a marvelous propensity to increase its diversity with the passage of time. This diversity is responsible for a range of global functions necessary for human survival, such as the biochemical flows of energy (through photosynthesis), water, carbon, nitrogen, and phosphorus, as well as providing a pool of genetic resources. Biodiversity offers other less utilitarian benefits as well—aesthetic (the beauty of so many living forms and their interaction) and intellectual (constituting a library of facts and relationships)."

We humans are finally becoming aware of the complex ways in which living things interact to maintain the delicate balance that sustains life itself. We know that even a little rupture in this complex interaction can create unforeseen and

Reprinted, by permission, from Bradley Shavit Artson, "Each After Their Own Kind," TIKKUN MAGAZINE, September/October 1997. TIKKUN MAGAZINE IS A BI-MONTHLY JEWISH CRITIQUE OF POLITICS, CULTURE, AND SOCIETY. Information and subscriptions are available from TIKKUN, 26 Fell St. San Francisco, CA 94102.

VRJC LIBRARY

harmful consequences elsewhere in the system. Our assault on the ozone layer, tropical coral reefs, and the Amazonian rain forest are but a few prominent examples. Hence, we are simultaneously agents and objects of our own newly-found (and little-comprehended) powers.

## The Jewish View

The matter of preserving biodiversity is of pressing Jewish concern. Since Judaism understands nature as God's creation, our religious experience at its foundation cultivates marvel at the teeming abundance of life and the diverse array of living things. Indeed, the experience of holiness through nature has inspired most of the world's religious traditions. Each faith tradition has responded to the wonder of the world in its own way; in Judaism, the preeminent response has been to see humanity as God's stewards, responsible to "guard and tend" creation (Gen. 2:15). As the psalmist reminds us, "the heavens belong to God, but the Earth was given to humanity," (Ps. 115:16). We have a special obligation to assure the continuing viability of creation, to maintain the Earth's bounteous ability to nurture life. Because Judaism recognizes humanity as God's stewards, our commitment to sustain diversity is nothing short of a religious mandate: a mitzvah.

Yet Judaism understands that people are both a part of creation and apart from it; we are little higher than the beasts, and "little lower than the angels," (Ps. 8:6). As unique creatures

> *"The matter of preserving biodiversity is of pressing Jewish concern."*

within creation, we can legitimately use the rest of creation to meet our needs as a species (just as all other animals do). As reflections of God's image in the world, however, we must also consider how well we are managing the world on behalf of its Creator.

While Judaism certainly allows people to use the resources of the world to sustain human development and well-being—permitting the taking of animals' lives for human nutrition and the harvesting of plants for human civilization— the Jewish balancing act that was established from the beginning requires us to "guard and tend" the garden in which we live, but do not own.

## The Story of Noah

To understand biodiversity within the world of Torah, one must look to biblical/rabbinic tradition and to its grand, sweeping stories of what creation means. Consider for example the story of the Flood and Noah's ark. "The earth had gone to ruin before God, the earth was filled with wrongdoing. God saw the earth, and here: it had gone to ruin, for all flesh had ruined its way upon the earth," (Gen. 6:11–12). Here the Torah deliberately updates itself: God's act of creation is very good, but the chaotic abuse by humanity has ruined it! Our actions have ruined the earth not only for ourselves, but also have imposed unwanted consequences on all living creatures: "all the residents of the world are governed by one and the same destiny," (Tanna deVei Eliahu Rabbah 2). In an attempt to restore creation, God resolves to send a flood, and instructs Noah, "a righteous, wholehearted man," to construct an ark: "From all [ritually] pure animals you are to take seven and seven [each], a male and his mate, and from all the animals that are not pure, two [each], a male and his mate, and also from the fowl of the heavens, seven and seven [each], male and female, to keep seed alive upon the face of all the earth," (Gen. 7:2–3). Such a command can only make sense if the survival of each and every species matters.

> *"The story of Noah's ark powerfully affirms the value of each existent species."*

Indeed, rabbinic tradition expresses this same message through a midrash in which the dove chastises Noah for endangering the survival of doves as a species: "You must hate me, for you did not choose from the species of which there are seven [in the ark], but from the species of which there are only two. If the power of the sun or the power of cold overwhelmed me, would not the world be lacking a species?" (Sanhedrin 108b). The story of Noah's ark powerfully affirms the value of each existent species, and highlights the role of humanity as God's partner in the preservation of biodiversity as it points to our ability to threaten that same variety.

## Jewish Law

The value of biodiversity finds repeated expression in the application of Halacha (Jewish law). Consider the following:

*Kilayim:* This category of Jewish law prohibits mixing diverse species together. It covers six kinds of mixed species: mixed seeds, grafting trees, seeds in a vineyard, cross-breeding animals, pulling cattle, and mixing linen and wool in garments (*shatnez*). Kilayim is prohibited by Halacha as an unwarranted tampering with the categories established by God's creation. In a similar vein, the Jerusalem Talmud understands the biblical verse "My statutes you shall keep," (Lev. 19:19) as referring to "the statutes I have engraved in the world," (Yerushalmi Kilayim 1:7). In other words, the laws of nature. Creation comes from God, which implies that the alteration of a natural law or the modification of a species constitutes an impermissible violation of creation. Ramban, the great medieval philosopher and sage, explains that "God has created in the world various species among all living things, both plants and moving creatures, and God gave them the power of reproduction, enabling them to exist forever as long as the Blessed God will desire the existence of the world," (Ramban referring to Lev. 19:19). The prohibition of kilayim is, therefore, an affirmation of species and diversity as they currently exist.

*Sending the Mother Bird Away:* The Torah records the insistence that one who gathers eggs from a nest must first shoo the mother away (Deut. 22:6). While the Torah doesn't reveal a reason for this practice, medieval rabbis were emphatic in linking this mitzvah to the preservation of species: In the Sefer Ha-Hinnukh, we are told that "God's desire is for the endurance of God's species . . . for under the watchful care of the One who lives and endures forever . . . it [every species] will find enduring existence through God," (*Sefer Ha-Hinnukh,* #515 of the 613 Mitzvot). Ramban speculates that "it may be that Scripture does not permit us to destroy a species altogether, although it permits slaughter [for food] within that group. Now, one who kills the dam and the young in one day, or takes them when they are free to fly, it is as though he cut off that species," (Ramban, to Deut. 22:6). Both of these authoritative rabbis understand this mitzvah as demonstrating the importance of maintaining each species of plant and animal. Jewish conduct must support the divine intention that each species thrives.

*Slaughtering the Animal and its Young:* Leviticus 22:28 prohibits slaughtering the mother ox or sheep and her young on the same day. As with the rule about releasing the mother bird, this law was understood in terms of assuring the continuation of existent species. In this area, Jewish thinkers articulate an explicit notion that providence extends over entire species. Just as environmental ethics values the species over the individual member of the species, Jewish thought insists that each species of animal has a "right" to exist that comes from, and is protected by, God. In the words of the *Sefer Ha-Hinnukh,* "One should reflect that the watchful care of the Blessed God extends to all the species of living creatures generally, and with God's providential concern for them they will endure permanently."

*Kashrut (Dietary Laws):* While the dietary laws don't speak directly to the issue of biodiversity, they do express the biblical emphasis on species as categories deserving attention and respect. In Leviticus 11 and Deuteronomy 14, the Torah delineates categories of animals which may and may not be eaten. One thoughtful scholar has suggested that the categories of *tahor* (pure) and *tamei* (impure) actually tell us what part of creation is available for our use (what we call "pure") and what part not authorized for human benefit (what we call "impure"). This remarkable observation highlights the fact that Judaism sees much of creation as existing for the satisfaction and self-expression of God, not humanity. Jews are not allowed to eat most animal species, and designating them as *tamei* reminds the observant Jew that the purpose of the world is not to please humans. As the midrash notes, "It is not thanks to you that rain falls, or that the sun shines—it is thanks to the animals," (Bereshit Rabbah 33:1).

> *"If God made all these species deliberately, we can do no less than assist in their continuing vitality."*

Each of these mitzvot direct Jews to demonstrate their reverence for creation as it is. Taken together, they form an essay on the subject of deeds that serve the

Creator by maintaining the creation as we find it. If God made all these species deliberately, we can do no less than assist in their continuing vitality. If God's loving care extends over the range of living things, our love must be sufficiently strong to keep them alive—as a tribute to our Creator, as the best defense for our own survival, and as an abiding expression of our love of life.

## Each Species Is Important

The beginning of Jewish values is the Beginning. Recognizing God as Creator, the Torah asserts that the world is not haphazard, coincidental, or meaningless. Rather, the expression of divine exuberance, as the embodiment and recipient of God's bounty, creation is saturated with meaning and value. The world's merit does not result merely from its usefulness to people, but from its status as God's creation. Each new species inspires the divine judgment "it is good" because the goodness of creation is enhanced by the addition of each new group. Consequently, each species has a vital role to play in the unfolding pageant that is life. We impoverish that drama each time we reduce the cast of characters. We literally diminish the greatness of God when we diminish the greatness of God's sovereignty.

In our role as stewards, we humans can preserve the diversity of living creatures. Embodied in forms ever more intricate and diverse, life becomes a polyphonic symphony in praise of its Source. The more the types of instruments, the more intriguing the melody; the more the number of species, the more resplendent the creation.

The Kabbalists saw biodiversity not in terms of a symphony at work, but of a congregation at prayer. In the words of Rabbi Nahman of Bratslav, "every blade of grass sings poetry to God," (Likkutei MoHaRaN, 306). Rabbi Abraham Isaac Kook affirmed that "everything that grows says something, every stone whispers some secret, all creation sings." For creation to function as a congregation, we must train ourselves to identify with all creatures and with all creation. We must discipline our desires and work for the well-being of the whole.

## A Vision of Unity

For the Jew, God's unity pervades creation: all are connected to each, and each to all. The unity of what God has created envelopes us, providing us with context and with goals. The diversity of all living things constitutes yet another context, a rainbow in which each species' distinctiveness adds to the resplendence of the totality, and in which any group's extinction impoverishes and endangers the rest.

At present, the oneness of all living things is hidden, masked by our species' self-regard and seeming self-reliance. But the illusion of humanity's dominion, maintained by our selfishness and delusions of power, will one day give way before the unification of God and God's name. After all, this vision of unity, of a time in which God, humanity, and creation live together in reverent balance is both as contemporary as environmental ethics and as ancient as Scripture. Our covenant, as Jews and as God's stewards, impels us to assure the flourishing of the diversity that God created. Our calling, as Jews and as caretakers of God's creatures, is to exert our best effort to protect all of life, and to fortify our faith in the bounteous flowering of life in which the Psalms rejoice:

*Praise God, sun and moon, praise God, all bright stars . . . !*

*Praise Adonai, all who are on the earth, all sea monsters and ocean depths, . . .*

*All mountains and hills, all fruit trees and cedars, all wild and tamed beasts, creeping things and winged birds, all monarchs and peoples of the earth!*

—Psalm 148:1–11

# Humans Should Not Be Indifferent Toward Other Species

## by Ed Ayres

**About the author:** *Ed Ayres is the editor of* World Watch, *a bimonthly environmental magazine published by the Worldwatch Institute.*

It may come as a shock to many to find that our closest genetic relatives on the planet—the primates—are diminishing in numbers at an alarming rate. If we were to draw a graph tracking the evolution of primates over the past four million years, this decline would appear, in the last 1-percent of the time-line, as a free-fall.

Primates are by no means the only category of life going into free-fall. Mammals in general are in decline; birds are in decline; amphibians are in decline; freshwater fish are in decline; and now we find that reptiles are in decline too. And, in numbers of species, all of these categories together add up to only a small fraction of the Earth's diminishing biodiversity. As Harvard biologist E.O. Wilson estimated a few years ago, thousands of species of smaller organisms are disappearing forever each year.

## Indifference Toward Wildlife

Not everyone who learns of this collapse is shocked by it, however. Many are indifferent; they don't see why it matters. And some are unapologetically hostile to any form of wildlife that interferes with human hegemony: witness the anger of U.S. ranchers toward coyotes, or of Zimbabwean farmers toward elephants, or of Japanese fruit growers toward monkeys. A few years ago, I spent some time in the Mojave Desert of California, where the Desert Tortoise is endangered. I learned that the single largest cause of tortoise death is bullets to the head, delivered by land owners who fear they will be prevented from "developing" the land by the tortoises' protected status. Ironically, the fact that they are

Reprinted, by permission, from Ed Ayres, "Outcompeting Ourselves," *World Watch*, September/October 1997.

"protected" seems to have made them *more* endangered.

Then there was the time I went out for a long run, up a road through the Angeles National Forest, and came upon a crew of workers in Forest Service uniforms clearing brush. Just as I passed, one of the men screamed. As I turned to look, he went berserk, slamming his machete at the ground, again and again, then leaping up and down, whirling, and wildly slashing every tree or bush within reach while bellowing obscenities at the top of his lungs. The other men stood gaping.

> *"The reason most primates other than we are in decline just as reptiles are, is that we are killing them."*

If, as I surmised, the man had been bitten by a rattlesnake, he was doing the worst thing possible—sending the venom racing into his heart. None of his fellow foresters suggested to him that he *calm down.*

I didn't stop to find out, but whatever snake had been disturbed in its habitat had doubtless been slashed to pieces. As I ran on, I wondered: if a professional custodian of our forests can be this alienated from the environment he's entrusted with "managing," how alienated—or just disconnected—are people at large? Most Californians weren't in the woods at all that day; they were in shopping malls, air-conditioned buildings, or their cars. But if I could have stopped them to ask, I suspect many would have said they wouldn't mind at all if the world's reptiles just disappeared. "Snakes? Crocodiles? Lizards? They *look* like dinosaurs, and so what if they go the *way* of dinosaurs?"

## The Loss of Primates

Reptiles have more importance to the stability of our environment than most of us realize—and may have more than we yet know about. After all, some of them have been around more than two-thousand times as long as *Homo sapiens* has. But if the fact that we're killing off some of our planet's most venerable examples of ecological success doesn't give us pause, how about the fact that we're also killing off our closest genetic relatives? We humans *are* primates, yet the reason most primates other than we are in decline just as reptiles are, is that we are killing them. Isn't *that* something people can relate to?

I hope so, but I'm not quite sure. It's no coincidence that the decline in primates has coincided with the explosion in human population. In biological terms, we're outcompeting our relatives; and as those who are hostile or indifferent to nature are prone to say, that's just the way evolution works. Competition, today, is almost a first principle of the dominant culture. When we were kids, we were told by athletic coaches that competition "builds character," and as adults we're told that it builds a healthy economy. It's Charles Darwin brought to the service of business. Not only are the intrinsic values of primates (not to mention their ecological roles) overlooked, but their genetic connection with us is disregarded.

What's most significant, perhaps, is that this reveling in conquest (or unthinking acceptance of it, at least) does not seem to stop at the borders separating our species from others. *Within* human society, too, competition is eroding diversity. Just as non-human primates are dying off, many non-dominant human cultures are dying off. Of the world's 6,000 distinct human cultures, some 3,000 are believed to be headed for extinction. Many of the remaining 3,000 are being weakened by the homogenizing influences of global media and commerce. Some analysts worry that we are moving too rapidly toward a world in which there will be only a few, highly dominant, cultures—or even just one.

## The Danger of Competition

I don't question that competition has great importance in human evolution, or indeed that it is deeply ingrained in our behavior and motivation. I have an irrepressible urge to compete, and generally get it out of my system by running marathons. I can be in 1,719th place in a race, half a kilometer from the finish, and will run my heart out to beat the stranger just ahead of me, so I can be 1,718th. There's no logic to it, but there's great emotional force—just as there was in the forester who went berserk, or in the ranchers who shoot tortoises. Competition, like technology, is neither good nor bad in itself. It can make a business more productive or more ruthless; it can make a species more robust or more rampant. Competition, whether economic or ecologic, is too powerful to be loosed on the world with no guidance or restraint.

For a long time, the victims of unrestrained competition were easily forgotten or marginalized; it's the winners who feast on Thanksgiving Day and control the writing of history books. But now we see a change happening: when there's too much losing, even the winners become victims. If human population continues to grow while the biodiversity of the planet continues to shrink, there will come a point (or points) where we will begin to regret our dominance. And the same regret now faces us as we extinguish cultural diversity. Just as an accelerating depletion of the gene pool begins to break down the web of life, the homogenization of cultures breaks down the diversity of creativity, inventiveness, perception, and sensibility that make up our collective global heritage.

# Efforts to Save Endangered Species Are Unfairly Criticized

## by Heather Abel

**About the author:** *Heather Abel is a writer for* High Country News, *a biweekly newspaper that reports on the West's public lands and natural resources.*

Idaho Rep. Helen Chenoweth stepped up to the podium at the Wise Use Leadership Conference in Reno, Nev., in the summer of 1995 and charged the Endangered Species Act with a series of assaults:

Californians lost homes to the 1993 fire because they were not allowed to clear weeds where endangered kangaroo rats live.

Snails smaller than a pencil point caused bankers to withhold loans— bankrupting Idaho ranchers.

Children may soon be ripped to shreds when the grizzly bear is introduced in Idaho, a state, she claims, it has never lived in.

Chances are, the Republican's audience of property-rights leaders had heard these stories before. The entire nation may know them since conservative Republicans have worked tirelessly to sell the American public on the idea that the Endangered Species Act declares war on private property.

### Bills in Congress

"Debate over the ESA has been fueled by anecdotes more than anything else in Congress," says Pam Eaton of the Wilderness Society, "and Republicans have successfully captured the debate." Proof of their success: A bill introduced by Reps. Richard Pombo, R-Calif., and Don Young, R-Ark., to eviscerate endangered species protection may come soon to the House floor. It would shield private property owners and pay them if environmental regulations decreased their property value. [The bill did not reach the House floor.]

Even though the bill has no less than 120 co-sponsors, passage in its current

Reprinted, by permission, from Heather Abel, "The Anecdotal War on Endangered Species Is Running Out of Steam," *High Country News*, November 13, 1995.

form looks less likely every day. Environmentalists and the U.S. Fish and Wildlife Service, working with the media, have chipped steadily away at the credibility of the horror stories. Polls continue to show strong public support for the law among both Democrats and Republicans, and in Congress, a rift has developed between moderate and conservative Republicans, forcing the party to aim for a more politically palatable bill.

> *"Coverage of [the Endangered Species Act] not only questions the horror stories, but highlights the many benefits plants and animals provide to humans."*

Chenoweth's anecdotes did not appear out of thin air. In 1992, several groups asked their members for personal stories of ESA abuse. The groups then took these stories—many of which agency officials acknowledge hold grains of truth—and churned out an entire folklore, according to Jim Jontz of Defenders of Wildlife. Organizations such as the Farm Bureau and the Timber Industry Labor Management Committee helped spread the word. By early 1994, the property rights movement was such a formidable force that the environmental community convinced its champions in Congress to delay attempts to reauthorize the Act.

## A Slow Reaction

But then followed the 1994 Republican landslide, which led Western Republicans to try to gut the law. The national press picked up the horror stories; both ABC's *20/20* and CBS's *Eye to Eye* ran exposés on rats rating more attention than people in the California fires.

Environmentalists reacted slowly. Some tried to counter with success stories of species recovery; most found their less sensationalized versions ignored by press and legislatures, says Hank Fischer of Defenders of Wildlife in Missoula, Mont. He charges that most journalists "mindlessly reported the horror stories."

The U.S. Fish and Wildlife Service, the agency responsible for implementing the Act, and the villain in most of the anecdotes, also seemed stuck in low gear. And Interior Secretary Bruce Babbitt appeared conciliatory, exempting landowners of five acres or less from the law.

But behind the scenes some within the agency fought back. Fischer says he'd get letters from staffers labeled "under private cover" and "for your information only." The letters showed the Fish and Wildlife Service could be "very effective at squashing the lies, but they were very reluctant to make a case themselves."

## Disproving the Stories

Other environmentalists received a plain white handbook from the agency without any insignia or author. The 26-page report, *Facts About the Endangered Species Act*, debunks 34 assaults on the Act. It reports, for example, that the General Accounting Office found no evidence that clearing weeds in kangaroo rat habitat would have saved houses from fire in California. Most horror stories, it explains, are either outright lies or overblown accounts of projects that were delayed temporarily by the law. The study says the Endangered Species Act stopped only four projects nationwide on private property between 1988 and 1992.

Fish and Wildlife Service Director Mollie Beattie finally appealed to the press. "The public needs help in making the connection between endangered species, their own health and welfare and their kids' future," she told the Society of Environmental Journalists Conference in May 1995. "This is an issue which we in the service have not defined well for the public. I know you can help."

Beattie's pleas may have worked. "Media coverage has changed drastically in the past six months," says Fish and Wildlife Service spokeswoman Megan Durham. "We now get many more phone calls from reporters wanting to know the truth." When the House Resource Committee held field hearings in the spring of 1995, many journalists reported that the panels of presenters were stacked against the law and some of the testimony overblown.

Recent coverage of the act not only questions the horror stories, but highlights the many benefits plants and animals provide to humans. In a recent editorial for the Cox News Service, Martha Ezzard tells the story of how a species of infection-fighting soil bacterium saved Senate Majority Leader Bob Dole's life after he was struck by an exploding shell in World War II.

## Cracks in the Opposition

The corporate backing of anti-ESA legislation has also become more apparent. Reporters jumped on a leaked memo that exposed Washington Republican Sen. Slade Gorton's ESA reform bill, which is similar to the Young-Pombo bill, as the handiwork of corporate lobbyists.

Some say the public's gullibility regarding scare stories was bound to come to an end. "It is extremely hard to mobilize grassroots opposition to an act when 99.9 percent of the American public have no negative experiences with it," says Michael Bean, attorney for the Environmental Defense Fund.

Cracks in the campaign to gut the ESA are also growing wider. In September 1995, two Eastern Republicans, Maryland Rep. Wayne Gilchrest and New Jersey Rep. Jim Saxton, introduced ESA bills which would continue to protect habitat on private land. These more moderate bills failed to make it through the House Resources Committee, but they struck a chord among the House leadership. . . .

## Misleading Moderation

Rep. Pombo insists support for an overhaul of the Endangered Species Act is still strong. "The Young-Pombo bill has 120 co-sponsors; no other bill has more than 10," says Pombo spokesman Matt Hartaman. "The numbers speak for themselves."

"Young and Pombo are pushing their bill fast now because they know their power will wane," responds Fischer.

Some environmentalists warn that the conservatives' campaign has skewed the debate so far to the right that new Republican concessions only *seem* moderate. "If Young-Pombo is an unmitigated disaster, then the Gilchrest bill can best be labeled a mitigated disaster," says Washington, D.C., attorney Eric Glitzenstein, who has brought dozens of lawsuits to protect endangered animals and plants.

Glitzenstein says many provisions of the bill—anonymous peer reviews of listings, non-federal teams to develop recovery plans and the requirement to minimize "adverse" social and economic consequences—could render protection impossible in the hands of a Secretary Watt "or even a Secretary Babbitt."

The appearance of moderation among some Republicans may be as calculated as the campaign of anecdotes, Glitzenstein warns. But Jontz celebrates the delay: "It's an indication that we've had an impact; we developed a division within the Republicans."

# Attempting to Save Every Species Is Expensive and Impractical

## by Ike C. Sugg

**About the author:** *Ike C. Sugg is a fellow in wildlife and land-use policy at the Competitive Enterprise Institute (CEI) in Washington, D.C. The CEI is an organization that supports the use of property rights and the private sector to protect the environment.*

When President Richard Nixon signed the Endangered Species Act in 1973, he inadvertently codified Aldo Leopold's stylish but stupid aphorism, "To keep every cog and wheel is the first precaution of intelligent tinkering." Should we lament the extinction of smallpox, or expand the ESA to protect unique life forms such as various viruses for future generations? Before we "Save the fungus among us!" as Gary Larson once quipped, we might want to reconsider what we are already trying to save—and, more importantly, how we are going about it.

### Expensive Bugs

As of May 1997, the ESA protects 33 species of insect. While most Americans might not think twice about stepping on a bug, the federal penalty if the bug turns out to be a rare one can amount to $200,000 in fines and one year in jail. And the same goes for modifying the habitat of such a species—even if the critters in question do not actually live on the land in question. In short, the ESA authorizes the Federal Government to prohibit *anything* that might encroach on or disturb species that are classified as "threatened" or "endangered."

Consider the case of the Delhi sands flower-loving fly, featured in an *NBC Nightly News* report on "The Fleecing of America." This fly is a native of the Colton Dunes in California's San Bernardino and Riverside Counties. It is known to live on less than 200 acres, all but 10 of which are privately owned.

Reprinted, by permission, from Ike C. Sugg, "Lord of the Flies," *National Review*, May 5, 1997. Copyright ©1997 by National Review, Inc., 215 Lexington Avenue, New York, NY 10016.

The fly spends most of its life underground, in a larval stage, and lives for only about a week above ground. During that short period of time, the males find mates and breed. After these encounters the females deposit their eggs beneath sandy soil. Then the adult flies die and the cycle begins anew.

NBC reported that the fly, listed as an "endangered" subspecies in 1993, has already cost the taxpayers of San Bernardino County some $4.5 million. The reason? Eight flies were seen buzzing around the building site

> *"Should we . . . expand the [Endangered Species Act] to protect unique life forms such as various viruses for future generations?"*

for a new hospital. The subspecies was listed less than 24 hours before construction was scheduled to begin; as a result, the county had to move the hospital, set aside almost ten acres of valuable land for a fly preserve, and monitor the fly's fortunes for the Federal Government. The county had paid approximately $400,000 per acre for that land; the Federal Government reimbursed none of that cost.

## The Interstate 10 Fiasco

As expensive as this was, it could have been worse. Initially, an official of the U.S. Fish & Wildlife Service, Linda Dawes, demanded that the county set aside the entire 68-acre hospital site for the fly. According to the sworn affidavit of a witness, Miss Dawes went so far as to demand that Interstate 10, an eight-lane freeway, be shut down—or traffic slowed to 15 mph—during August and September, when the flies emerge for their week-long sojourn above ground. She worried that one of these rare bugs might end up on the windshield of a speeding car.

In the end, construction of the hospital went forward, I-10 remained open, and Linda Dawes left her job at Fish & Wildlife. But that does not mean that all is well in San Bernardino County. The intersection of Valley Boulevard and Pepper Avenue near the hospital's main entrance is severely congested. Unless the approach to the hospital is improved, a recent study concluded, "virtual gridlock" will result when the hospital opens. Emergency access will be slow, costing precious time to those for whom a few minutes can make the difference between life and death.

The Fish & Wildlife Service, however, insists that the county's plan to reconfigure the intersection will "encroach" on a "migration corridor" for the fly, thus "harming" the fly and violating the ESA. No flies have ever been documented as using this migration corridor, but that doesn't matter. The Service has a theory.

According to Service biologists, the flies need to be able to travel, unimpeded, over natural vegetation, from the hospital preserve to suitable habitat about a quarter-mile away. Presumably, the flies will head west, following the

100-foot-wide corridor designed for them, then make a 90-degree turn to the north, where the flyway narrows to 30 feet, and follow that path for 700 feet. At the end of the flyway, the flies will make a 90-degree turn to the west and cross a four-lane street, on the other side of which their habitat awaits.

Assuming that the flies will understand and follow the flight path laid out for them, one wonders how they will make it across the busy street. Common sense suggests that oncoming traffic would pose a greater threat than an overly narrow flyway. The county proposed to retain an 18-foot-wide migration corridor for the fly, but the Service said that the proposal was "not biologically justified" and threatened to sue the county if it went forward without "mitigating" for the flyway. In other words, the county will have to buy the Delhi sands flower-loving fly yet another preserve.

## Increased Business Costs

It gets worse. These imperiled insects live alongside imperiled people, in an area of historically high unemployment and economic distress. To attract industry and create jobs, the 10,000-acre Agua Mansa Enterprise Zone was established in 1986, one of California's first enterprise zones. It offers specific tax advantages to businesses that locate within the zone. But there is a fly in the ointment.

> *"Extortion . . . is commonplace under the ESA."*

For CanFibre, Inc., a Canadian manufacturer of recycled fiberboard, the Delhi sands flower-loving fly has added $450,000 to the cost of building a plant in the Agua Mansa. The Fish & Wildlife Service claimed that CanFibre's entire 300-acre property was occupied by the fly, and that the company would have to give up 65 acres to "mitigate." Tom Olsen, a biologist who consulted on the project, says the government's demands rest on "rhetorical horse manure."

"Less then 10 acres of suitable habitat exists on the property," says Olsen. Indeed, his biologists "had [only] two sightings of males flying over the property; there was no evidence of use by females, and the two males we identified could have been the same fly." As with the hospital, the feds backed off on halting the project, but they collected a cool $450,000 in cash.

## The Drawbacks of Habitat Conservation

Such extortion—wherein landowners fork over money, land, or both in exchange for permission to use their own land—is commonplace under the ESA. Indeed, that's what "habitat conservation plans" are all about. The cost of obtaining such permission has deterred at least one firm, Trism, Inc., from locating in the zone. "What they want is a fly park," said a Trism representative.

Since then, the Fish & Wildlife Service has come up with a new idea: it has announced that it wants to set aside 200 to 300 acres in the enterprise zone for yet *another* fly preserve (that makes three, for those who are keeping count).

The Service claims this preserve will actually make things easier for the area's beleaguered landowners. Rather than forcing individual property owners and developers to run the regulatory gauntlet for development permits one at a time—an expensive and time-consuming process—the Service wants to create one grand extortion scheme through which landowners pay collectively.

Such region-wide habitat conservation plans are supposed to be good for everyone. The fly gets a big preserve, and the landowners get "regulatory certainty." Since the fly preserve will be clearly defined, landowners will know up front whether the government is going to take their land. Those who own land nearby will pay a "development mitigation fee" to help compensate their neighbors whose land is to be included in the preserve. In other words, those who have flies on their land will be dispossessed, while those who do not have flies on their land will be robbed. Despite the admonition in the Fifth Amendment: "nor shall private property be taken for public use, without just compensation," the Federal Government is here proposing not to distribute this cost across the whole public through taxation, but rather to impose it on a handful of people who happen to own neighboring property.

Local planners say the preserve land is worth approximately $35,000 per acre. That means a 300-acre fly preserve will cost $10.5 million—to be paid by those who develop land that is completely devoid of endangered flies. Because the ESA requires such "mitigation" to be paid only when "actual death or injury" occurs to protected wildlife, and because those who will be paying the mitigation do not have such wildlife on their land, the government's extortion scheme is patently contrary not only to the Constitution but also to the Act it is supposedly enforcing.

## Controlling Development

Many landowners will capitulate anyway. For those with liens on their property and notes to pay, fighting the feds is a losing proposition. In their view, it is a choice of being allowed to develop some land or none at all. Environmentalists claim such plans are totally voluntary, but in reality they present a Hobson's choice.

Even if developers buy into the scheme, it is likely to take them years to come up with the millions necessary to pay for the preserve. And even if the dispossessed landowners eventually get paid $35,000 an acre for their land, that may be far less than what it is really worth. San Bernardino County paid between $150,000 and $400,000 per acre for the nearby hospital site. As luck would have it, the most expensive land was where the eight flies were found. Now the flies are the de facto owners of prime real estate, taken on their behalf by the Federal Government.

> *"Controlling development, rather than saving endangered species, has become the point of the ESA."*

While some might argue that the ends justify the means, in this case it appears that the ESA is causing all this trouble for nothing. "The extinction of the Delhi sands flower-loving fly in the immediate future is a likely event," according to the Interior Department's Draft Recovery Plan for the fly. The authors of the 1996 Recovery Plan predicted that the fly would go extinct even with the protection of the ESA's land-use regulations.

So why all the regulations? The answer is clear: Controlling development, rather than saving endangered species, has become the point of the ESA. Indeed, if environmentalists were honest they would admit that the ESA is encouraging habitat destruction, not conservation.

Endangered species are the last things landowners want on their property, and many are taking affirmative actions to ensure that their land is devoid of such species. This is the ESA's dirty little secret. The only way to correct such perverse incentives is by explicitly removing the Federal Government's claimed authority to take private property without paying for it. Until Congress does this, the United States' government-run protection racket will continue to wreak havoc on the lives of its hapless victims, both human and non-human.

# Belief in Biodiversity Is Dangerous

## by Alston Chase

**About the author:** *Alston Chase is a nationally syndicated columnist on the environment.*

First, the news: The Endangered Species juggernaut continues to roll. On Sept. 12, 1994, the National Marine Fisheries Service announced it would consider declaring as endangered "all salmon and anadromous [i.e., sea-run] trout populations in Washington, Oregon and California." Three days later, the U.S. Fish and Wildlife Service listed four species of fairy shrimp in California's Central Valley as endangered.

Now the question: What is the historical significance of these events? We already know the Endangered Species Act is a scientific fraud and economic calamity. It is designed to maintain a balance of nature that never existed. It pursues the impossible dream of halting evolution. Its partisans say it promotes "biodiversity," but they won't define the term. It spreads unemployment and costs the country billions.

## The Perceived Power of Genes

But it is also another sign that dangerous notions of biological determinism are in vogue. *Time* magazine avers that marital infidelity "may be in our genes." The courts increasingly favor dysfunctional natural parents over loving adoptive ones; feminists argue that gender taints everything from scholarship to social behavior. And according to the September 1994 issue of the *Atlantic Monthly*, the once-taboo idea that heredity controls behavior is back in style.

Indeed, this fascination with genetic imperatives is largely a product of America's obsession with ecology. The central insight of this science is that all things are interconnected within "ecosystems" so that the disappearance of one creature undermines the survival of all others. Thus, it is supposed, loss of biodiversity threatens civilization.

Alston Chase, "Poison Pill of Biodiversity," *Washington Times*, September 13, 1994. Reprinted with permission from the author and Creators Syndicate.

## Chapter 2

### Biology and Nazism

But long ago this idea proved to be the poison pill of politics. Conceived by Prussian botanist Ernst Haeckel in 1866, ecology quickly led to unsavory insights. Since all things form a seamless web, Haeckel reasoned, there are no fundamental differences between people and other creatures. Hence, he rejected humanism, Judaism and Christianity, which conferred importance to individuals. Believing Germany weak because these ideas separated people from nature, he theorized that re-establishing links with the land demanded reviving original "Volkish" culture.

Popularizing ecology until it became what historian Anna Bramwell dubbed the "German disease," Haeckel was the founding father of Nazism. Jews, deemed to embody anti-nature values such as technology, science and consumerism, were seen as infecting the body politic. "The evidence is ample," Miss Bramwell noted, that "there was a strain of ecological ideas among Nazis." Racialism, philosopher Karl Potter observed, "can be traced back to Haeckel." In his book, *The Nazi Doctors*, Robert Jay Lifton concluded Haeckel "undoubtedly inspired Hitler and other high-ranking Nazis." Historian Daniel Gasman called Haeckel "Germany's major prophet of political biology." And German historian H. Schnadelbach wrote, "the general biologism in the theory of culture . . . culminated in National Socialism."

> *"The Endangered Species Act . . . is designed to maintain a balance of nature that never existed."*

National Socialism, Hitler's deputy Rudolf Hess explained, was "applied biology" intended to restore the "vitality of the German race." It sought "biological renewal," said another Nazi, Werner Best, through building an "organically indivisible national community." And those opposed were "the symptom of an illness which threatens the healthy unity of the . . . national organism."

"The Nazi ethos," Mr. Lifton concluded, "thus came to contain a sacred biology" that led directly to Auschwitz. "The Nazis saw themselves as practitioners of the science of life and life processes as biologically ordained guides to their own and the world's biological destiny."

While killing Jews, the Third Reich simultaneously sought to preserve forests and wildlife, which symbolized the nation's Volkish past. In 1933, it launched a "Back to the Land" program to create a new rural Germany. Subdivisions were declared illegal; properties confiscated, sanctuaries created, and hedge-row and copse protection ordinances passed "to protect the habitat of wildlife." Hess promoted "bio-dynamic farming," and SS leader Heinrich Himmler established organic farms, including one at the Dachau concentration camp where inmates grew herbs for SS medicines.

Having forgotten this history, Americans may be condemned to repeat it. Unaware of the past consequences of their ideas, environmentalists condemn humanism, Judeo-Christianity, capitalism, materialism, private property, technology and consumerism, even as they revere nature, primitive culture, organic farming and wildlife reserves.

Spending millions to eradicate "exotic" species, preservation agencies manifest an obsession with biological purity. The Endangered Species Act applies racialist theory to plants and animals. Seeking to shield the genetic purity of "native" creatures, it prohibits such steps as rearing salmon in hatcheries or feeding grizzly bears, for fear this might pollute "wild" genes with human-induced mutations.

Of course, no American would knowingly condone atrocities in the name of protecting ecologic health. Nevertheless, by flirting with the flawed idea that biology is destiny, the nation courts political suicide. Subordinating individual welfare to vague abstractions such as "biodiversity" not only threatens liberty but, more importantly, underestimates the human spirit. In the end, survival depends less on material circumstances than on our ability to transcend them.

# Attempts to Preserve Species Endanger Human Safety and Property

**by Roger B. Canfield**

**About the author:** *Roger B. Canfield is a writer for the* New American, *a biweekly magazine published by the John Birch Society, a conservative organization that advocates the abolition of many federal regulatory agencies.*

In the spring of 1997, as in the previous few years, major flooding around the U.S. has demonstrated that while man is given stewardship over the vast and varied earth God has created, he is nonetheless at the mercy of the elements when conditions grow extreme. While man is required to plan and implement strategies for managing his land resources—and does so with remarkable ingenuity and efficiency—on occasion those strategies prove inadequate to the terrible forces of nature.

Witness the devastating floods which have wrought untold sorrow, death, and destruction on those living in the Red River Valley of North Dakota. The flood, the type which experts say hits a given area perhaps once every 500 years, defied flood control systems that had been put in place over the years, causing an estimated $1 billion in damages. While there is some proof that flood-control programs in the most devastated areas were not as aggressive as they might have been (at the time of the flood Grand Forks was in the planning phase of a $40 million project to complete its partial levee system), experts agree that North Dakota's "Flood of '97" would have defied even man's most conscientious efforts to reject its fury.

## The Prevention of Flood-Control Projects

Unfortunately, some of the spring flooding in recent years has been exacerbated by environmental regulations which have delayed or eliminated necessary flood-control measures. In particular, through the federal Endangered Species

Reprinted, by permission, from Roger B. Canfield, "Saving People or Species," *The New American*, June 9, 1997.

Act (ESA), the "protection" of various "endangered" species of plants and wildlife has halted vital flood-control projects that would have saved lives and billions of dollars in property in California and along the Missouri and Mississippi Rivers during flooding from 1993–97.

The U.S. Fish and Wildlife Service (FWS) and state and local governments enforcing the ESA actively obstruct day-to-day flood control work—all for the stated purposes of protecting the habitat of such "endangered" species as the fan shell mussel, the rough pigtoe mussel, and the white warty back mussel, to name just a few. For several years, the clearing of weeds, trees, and silt from flood channels has been halted by wildlife agents, severely compromising the structural integrity of earthen levees and clogging channels.

> *"Some of the spring flooding ... has been exacerbated by environmental regulations which have delayed or eliminated necessary flood-control measures."*

Two California Republican congressmen, Representatives Richard Pombo and Wally Herger, attempted to address the ESA obstacle through H.R. 478, the Flood Prevention and Family Protection Act. While the bill, which would have exempted current flood control projects from pertinent ESA restraints, was effectively gutted in House floor action on May 7, 1997, and then withdrawn from consideration, it points out that there are key conservative lawmakers who are concerned about the problem.

Steve Thompson, spokesman for Representative Herger, said that in California, over 30 levees failed during winter and spring flooding in 1997. These floods killed nine persons, displaced 100,000 people, and caused $1.6 billion in damages to public and private property. According to Herger's office, the losses were worsened due to levee repair projects delayed since 1991.

## ESA-Enforced Delays

Recently Representative Pombo, a strong advocate of ESA reform and property rights, has brought California flood-control experts before the House Resources Committee. His witnesses flatly blamed part of the California flood losses on the ESA. They reported overrun riverbanks, clogged tributaries, and broken levees along the Sacramento, Feather, Bear, Consumnes, and San Joaquin Rivers. Major flood damage could be found along a nearly 300-mile river basin front north and south of the Delta drain into San Francisco Bay. Levee experts said that a good share of the damage could have been prevented had it not been for years of ESA-enforced neglect of routine flood control maintenance, which led to overgrown rivers and compromised levees.

In Pombo's district, years of accumulated silt and weeds clogged up the San Joaquin River basin. Alex Hildebrand, a director of both the South Delta Water Agency and the California Central Valley Flood Control Association, explained the problem: "We need to take the sediment out of the bottom of the river and put it up on top to thicken the levees. . . . They [EPA enforcers] won't allow you. . . . The river bottoms have come up about eight or nine feet . . . they haven't been dredged in years."

Similarly, in early 1995, floods along the Pajaro River caused $240 million in damages in Monterey County alone. California State Assemblyman Peter Frusetta blamed the flooding on seven years of ESA delays in Pajaro riverbed maintenance. Despite denials by environmentalists, the Monterey Grand Jury declared that unkempt rivers were "a very dominant" force in the 1995 flood. Phil Larson of the Fresno Farm Bureau charged that ESA restrictions on silt removal and levee maintenance led to a flash flood washout of a bridge under California's Interstate 5 that killed eight people in 1995.

Recognizing the need for some relief for flood-control repairs, in 1995 California Governor Pete Wilson ordered state enforcers of ESA measures to back off and allow "general permits" to be issued for needed maintenance and repairs. In early 1997 he formed the Flood Emergency Action Team (FEAT), headed by Douglas Wheeler, State Secretary of Resources. But in the February 1997 issue of *Inside California,* investigative writer Sarah Foster noted that Wheeler had previously worked for the World Wildlife Fund and the Sierra Club and thus could not be expected to "choose flood control measures over wildlife habitat." Nonetheless, Wilson's March FEAT report claims that the State Department of Fish and Game (DFG) was on the

> *"Bureaucratic obstruction of vital flood-control works is done in the name of some very peculiar critters and vegetation."*

1997 flood scene expediting cleanup and repairs. Concerning the levees, the DFG determined which "mitigations" were necessary to protect species and issued permits to repair levees.

Unlike the DFG, the federal FWS was, noted the FEAT report, "not issuing permits and [was] deferring all mitigation. . . . [T]here is no clear direction from the USFWS as to what . . . mitigation will be required or when. . . . [T]his could affect the speed and . . . ability . . . to complete the repairs. . . . [T]he USFWS may require mitigation for the emergency repairs. [T]his . . . will reduce the funds available for [actual] levee repair."

## Protecting Peculiar Species

Bureaucratic obstruction of vital flood-control works is done in the name of some very peculiar critters and vegetation protected under the ESA and conveniently found at sites in desperate need of repair. These "endangered" species include:

• *Parrot Feather Weed.* Sarah Foster reported that near Gridley, California, a state DFG game warden arrested a backhoe operator for clearing weeds from a drainage ditch without a DFG permit. Finding the noxious parrot feather weed during a rainy November, the local Reclamation District 777 assumed they needed no permit under the "general permit" order. "It's crazy," Robert Millington, an attorney in the case, told Foster. "Either we're going to maintain the flood control . . . as we have since 1907 . . . or [the DFG] will turn this valley into a wildlife habitat."

• *Giant Garter Snake.* Claiming habitat for garter snakes near Robbins, California, DFG agents ordered a halt to work on a weakened levee. The DFG did not want to disturb hibernating snakes. According to Representative Wally Herger, no snakes had been seen in the area.

• *Delta Smelt.* Dante John Nomellini, manager of California's Central Delta Water Agency, said that the ESA "prohibition of dredging and . . . of fill for levee maintenance and the creation of shaded riverine aquatic or emergent marsh habitat . . . for delta smelt" created flood losses in 1997 in the Sacramento Delta.

• *Elderberry Bushes and Elderberry Beetles.* There are "large numbers of elderberry bushes and their . . . protection . . . for a limited number of endangered elderberry beetles is . . . an abuse of the ESA," wrote Nomellini. Bushes and beetles were an ESA priority throughout most of California's vast Central Valley as early as 1993.

• *Long-toed Salamander and Steelhead Trout.* California Assemblyman Peter Frusetta notes that "the Pajaro River being designated . . . habitat for the long-toed salamander and steelhead trout made it impossible to clear the river of excess trees, vegetation, and debris, causing excess flooding."

## Unsafe Levees

After their Missouri constituents faced ESA-enhanced flood damage along the Mississippi and the Missouri Rivers, Republican Missouri Congressmen Kenny Hulshof and Jo Ann Emerson signed on to H.R. 478. Hulshof's office said that levee repairs in Missouri were delayed many months after the 1993 Missouri River floods. Carl Lensing, a farmer and the founder of the Missouri River Levee and Drainage District Association, said that the delays took some eight months. Environmentalist officials insisted upon preserving trees for eagle nesting. "Nesting birds outrank people along the Missouri," Lensing mused. Afterwards, they saved potholes caused by rushing floodwaters and preserved "borrow" pits from which earth had been removed to shore up levees—new spawning locations for minnows, crawdads, and "mud bugs."

> "*Of course, there are those 'environmentally correct' souls who fight any efforts to protect people and property.*"

David LaValle, spokesman for Representative Emerson, said that the ESA continued to "slow down repairs of levee structures" even after the Mississippi floods of 1993 and 1995.

George Grugett of the Lower Mississippi Valley Flood Control Association, which represents private and public flood organizations from Illinois to the Gulf of Mexico, detailed ESA protections of "endangered" species during levee repair. One species was the Fat Pocket-book Pearly Mussel. According to Grugett, "[S]hortly after work began a dead mussel . . . was discovered. . . . Since [it] was . . . listed as endangered, [dredging] work . . . stopped."

## Bears and Eagles

Another "protected" species is the Louisiana black bear. The FWS proposes to designate three million acres in Louisiana and Mississippi as critical habitat for the Louisiana black bear. Grugett predicts that "we will be hard-pressed to bring those 300-plus miles of deficit levees . . . to the required grade and section. When those levees fail . . . not only will the Louisiana black bear be in critical danger, but so will about 4 million people. . . ."

Gary Heldt of the Missouri Farmers Association recounted the efforts to "protect" Bald Eagles: "We had to get all together down on the river bottoms where we inspected each tree—determining which tree to cut and to save." Trees four inches in diameter at head height were saved, just in case "an eagle on the ESA list might fly by and decide to nest there." Tom Waters of the Missouri Levee and Drainage District said that environmental regulations protecting the eagle not only caused costly delays in flood control, but were part of a "land grab" agenda. Carl Lensing speculated that the whole scheme is a "fraud, using bug, fish, or worm to convert private property to public use."

## The Flood-Control Battles

H.R. 478 would have bypassed relevant provisions of the ESA whenever "building, operating, maintaining or repairing" any current flood control measure was designed to prevent an imminent threat to public safety or a catastrophic natural event. Also outside the reach of the ESA was any "routine operation, maintenance, rehabilitation, repair or replacement" of a flood-control project which had previously passed federal muster. Passage of the measure might have cut back the magnitude of ESA meddling and mayhem. The FWS alone had consulted on 18,211 federal government projects by 1994, according to the General Accounting Office—not counting local and private projects.

Of course, there are those "environmentally correct" souls who fight any efforts to protect people and property. California State Senator Tom Hayden, chairman of the state senate Committee on Natural Resources, would like to see the Los Angeles River restored to its natural state—complete with droughts and flash floods. Interior Secretary Bruce Babbitt just wants to blow up dams. The American River Association wishes to return the Missouri River to its natural state—what Missouri farmer Carl Lensing describes as "a meandering morass of muddy swamps and sloughs."

In 1996–97, new wetlands regulations gave the U.S. Army Corps of Engineers expanded environmental authority which could well impede both recovery from the 1997 floods and preparations for future floods. The Nationwide Permit Program requires the Corps to "minimize adverse effects on the aquatic environment [to] ensure that endangered species and their habitat are fully protected." It also imposes increased reporting, notification, and review requirements on persons seeking development permits in wetlands. The Water Resources Development Act of 1996 requires the Corps to restore environment degraded by Corps projects, to build small aquatic ecosystems, to plan improvements of watersheds and ecosystems, and to develop "nonstructural flood control technologies."

Any legislative efforts, such as H.R. 478, face tough opposition from environmental extremists plugged into key positions in the Clinton Administration and other government and quasi-government agencies. It is a classic battle between the vast collectivist army of environmental bureaucrats and activists and the champions of private property and individual freedoms. May God aid us in the fight.

# Environmentalists Overstate the Importance of Certain Species

**by David Andrew Price**

**About the author:** *David Andrew Price is a writer and an attorney with the Washington Legal Foundation, an organization that seeks to shape public policy and defend free enterprise and individual rights.*

One morning in January 1994, Arvid Enghaugen, a resident of the Norwegian coastal town of Gressvik, found his whaling boat sitting unusually deep in the water. When he climbed aboard to investigate, he found that the ship was in fact sinking; someone had opened its sea cock and padlocked the engine-room door. After breaking the lock, Enghaugen discovered that the engine was underwater. He also found a calling card from the Sea Shepherd Conservation Society, a small, California-based environmentalist group that specializes in direct actions against whalers. Counting Enghaugen's boat, Sea Shepherd has sunk or damaged eleven Norwegian, Icelandic, Spanish, and Portuguese vessels since 1979.

## Greenpeace Versus Whalers

The boat was repaired in time for the 1994 whaling season, but Enghaugen's problems weren't over. On July 1, 1994, while he was looking for whales off the Danish coast, five Greenpeace protesters boarded the ship from an inflatable dinghy and tried to take its harpoon cannon. Enghaugen's crew tossed one protester into the sea, and the rest then jumped overboard; the protesters were picked up by the dinghy and returned to the Greenpeace mother ship.

A week later, after Enghaugen's boat shot a harpoon into a whale, a team from another Greenpeace vessel cut the harpoon line to free the wounded animal. A group again tried to board the whaler, and the crew again threw them off. Enghaugen cut a hole in one of the Greenpeace dinghies with a whale flensing knife. For the next two weeks, Enghaugen and crew were dogged by Greenpeace ships and helicopters.

David Andrew Price, "Save the Whalers," *The American Spectator*, February 1995. Copyright © The American Spectator, reprinted by permission.

Although the activists failed to stop Enghaugen's hunt, their public relations war in America has been a different story. Over the past twenty years, the save-the-whales movement has been so successful in shaping public sentiment about the whaling industry that the U.S. and other nations have adopted a worldwide moratorium on whaling. Part of the credit must go to the animals themselves, which are more charismatic on television than Kurds, Bosnians, or Rwandans, who have engendered far less international protection. The movement owes most of its success, however, to the gullibility of Hollywood and the press in passing along bogus claims from whaling's opponents.

## The Endangered Whale Myth

The mainstay of the case against whaling—that it threatens an endangered species—is characteristic of the misinformation. It is true that European nations and the United States killed enormous numbers of whales during commercial whaling's heyday in the nineteenth century, but to say that "whales" are endangered is no more meaningful than to say that "birds" are endangered; there are more than seventy species of whales, and their numbers vary dramatically. Some are endangered, some are not. The blue whale, the gray whale, and the humpback were indeed depleted, but those species were later protected by international agreement long before the existence of Greenpeace or Sea Shepherd. (There have been abuses. Alexei V. Yablokov, special adviser to the president of Russia for ecology and health, has revealed that the whaling fleet of the former Soviet Union illegally killed more than 700 protected right whales during the 1960s, but the International Whaling Commission's (IWC) institution of an observer program in 1972 essentially put an end to the Soviet fleet's illegal activities.)

The only whale species that Enghaugen and his fellow Norwegian whalers hunt is the minke, which Norwegians eat as whale steaks, whale meatballs, and whaleburgers. As it turns out, minke whales are no more in danger of extinction than Angus cattle. In 1994, thirty-two Norwegian boats killed a total of 279 minkes, out of an estimated local population of about 87,000 and a world population of around 900,000.

In 1982 the IWC voted to suspend commercial whaling for a five-year period starting in 1986. The ostensible purpose was to permit the collection of better data on whales before hunting resumed. Norway lodged a reservation exempting itself from the moratorium, as the IWC treaty permitted, but it complied voluntarily.

Whaling nations soon learned, though, that the majority of nations

> *"To say that 'whales' are endangered is no more meaningful than to say that 'birds' are endangered."*

in the IWC—including the United States—intended to maintain the ban indefinitely, no matter what the numbers showed. Canada left the IWC in 1982, and Iceland left in 1992. Norway terminated its voluntary compliance in 1993. To protest the commission's disregard of the facts about whale stocks, the British chairman of the IWC's scientific committee resigned that year, pointing out in his angry letter of resignation that the commission's actions "were nothing to do with science." The IWC continued the moratorium anyway at its next meeting.

> *"Popular notions of whales' humanlike intelligence, often cited by opponents of whaling, have little real support."*

A 1993 report by the Congressional Research Service observed that the data on whales undercut the conservationist argument, and that "if the United States argues for continuing the moratorium on commercial whaling, it may have to rely increasingly on moral and ethical appeals." The ban on whaling is no longer about conservation, in other words, but about the desire of many Americans and Western Europeans to impose their feelings about whales upon the whaling nations (which include Iceland, Russia, Japan, and the Inuits of Canada and Alaska).

## No Exceptional Intelligence

Popular notions of whales' humanlike intelligence, often cited by opponents of whaling, have little real support. Whales possess large brains, but that proves nothing about their mental agility. Margaret Klinowska, a Cambridge University expert on cetacean intelligence, holds that the structure of the whale brain has more in common with that of comparatively primitive mammals such as hedgehogs and bats than with the brains of primates.

Whales can be trained to perform stunts and other tasks, but so can pigeons and many other animals that have never been credited with the cerebral powers of *homo sapiens.* And the idea that whales have something like a human language is, at present, pure folklore. Like virtually all animals, whales make vocalizations, but there is no evidence that they are uttering Whalish words and sentences. Their famed "singing" is done only by the males, and then during but half the year—a pattern more suggestive of bird-song than human speech.

Much of the popular mythology about cetacean intelligence comes from crank scientist John Lilly, a physician who became convinced in the 1950s that whales and dolphins are not only smarter and more communicative than humans, but also have their own civilizations, complete with philosophy, history, and science that are passed down orally through the generations. His conclusions about the animals' mental skills were based partly on his observations of captive dolphins at his lab in the Virgin Islands, but mainly on wild flights of conjecture. Lilly also predicted in the late seventies that the State Department would eventually negotiate treaties with the cetaceans, and that humanity's progress in its dealings with them would lead the Galactic Coincidence Control Center to send agents to planet Earth to open the way for extraterrestrial contacts with us. The anthropomorphization of the whale reached new heights with a 1993 open letter to the Norwegian people from Sea Shepherd president Paul Watson, who predicted, "The whales will talk about you in the same vein as Jews now talk of Nazis. For in the eyes of whalekind, there is little difference between the behavior of the monsters of the Reich and the monsters behind the harpoon."

## Cetaceans Are Not Humans

Cetacean behavior researchers have rejected Lilly's claims. Dolphin investigator Kenneth Norris of the University of California Santa Cruz, who was among the first to study dolphins in the wild and is responsible for much of our knowledge about dolphin sonar, writes that they have "a complicated animal communication system, yes, but for an abstract syntactic language like ours, no compelling evidence seemed, or seems, to exist." The late David and Melba Caldwell, who studied dolphin behavior at the University of Florida, maintained flatly that "dolphins do not talk." In their view, "dolphins probably are just exceptionally amiable mammals with an intelligence now considered by most workers, on a subjective basis, to be comparable to that of a better-than-average dog."

Louis Herman, director of the University of Hawaii's marine mammal laboratory and an opponent of whaling, has been studying the behavior of captive dolphins since 1967 and of whales in the wild since 1976. Herman says he has seen no evidence that the natural vocalizations of dolphins constitute a language. And for whales? "There's no reason to think the situation would be different with other cetacean species," he answers.

What American policy on whaling enforces is simply a cultural preference—one comparable to our distaste for horsemeat, which is favored in France. The whale-savers have succeeded in shaping policy by selling the idea that whales are different: that they are endangered underwater Einsteins. That's why Icelandic filmmaker Magnus Gudmundsson, who has produced a documentary showing Greenpeace's machinations on the issue, is correct in calling the movement "a massive industry of deception."

# Chapter 3

# How Can Pollution Be Reduced?

# Chapter Preface

In the novel *Don Quixote*, the protagonist fights windmills, believing them to be giants. In modern society, windmills and wind power are considered by some people to be friend, not foe. These people believe wind power may be one way to reduce air pollution.

According to its supporters, wind is an inexpensive, plentiful, and nonpolluting source of energy. Wind power advocates contend that wind power is not much more expensive than fossil fuel-based electricity; producing electricity from natural gas costs three cents per kilowatt-hour, while wind-generated electricity costs five to seven cents per kilowatt-hour. Advocates also note that unlike fossil fuel, wind is not a finite resource. According to Tracey C. Rembert, "The contiguous U.S. has enough untapped wind energy to produce 4.4 trillion kilowatt-hours of electricity a year—more than one and a half times our total electricity generation in 1990."

Opponents of wind power maintain that relying on wind as a key source of energy would be, in effect, a quixotic effort. For example, Robert L. Bradley Jr., president of the Institute for Energy Research, argues that wind power is expensive and impractical. He contends that wind generates electricity at low capacity due to the fact that wind does not blow twenty-four hours a day. "Because wind is an intermittent (unpredictable) generation source, it has less economic value than fuel sources that can deliver a steady, predictable source of electricity." According to Bradley, wind power also is noisy, uses up land, and even poses a threat to nature—wind blades have reportedly killed thousands of birds in the 1990s.

Wind power is just one approach to reducing pollution. In the following chapter, the authors debate whether air and water pollution need to be reduced and, if so, what methods can best solve the problem.

# Stronger Air-Pollution Standards Are Needed

**by Jeffrey St. Clair**

**About the author:** *Jeffrey St. Clair is the environmental editor of* Counter-punch*, a bimonthly political newsletter.*

When Bill Clinton journeyed to the north rim of the Grand Canyon in the fall of 1996 to preside over the creation of a new national monument, he quipped to reporters that it was kind of odd that there was so much fog in Arizona at that time of year. That wasn't fog, Mr. President, it was smog, clogging the air in one of the most remote and least populated areas in North America. The pollution shrouding the Grand Canyon had wafted from the smokestacks of coal-fired power plants and refineries hundreds of miles away.

## Increasingly Toxic Air

The Grand Canyon is only one of dozens of national parks where toxic compounds in the air are stunting tree growth and killing alpine flora. The situation in America's cities is even worse. More than 400 counties now exceed federal air-pollution levels. No fewer than 185 different scientific studies support the conclusion that the nation's air is becoming ever more toxic.

Nearly 130 million Americans are exposed every day to harmful levels of air pollution. Studies at the Harvard Medical School estimate that 60,000 Americans die prematurely every year from respiratory illnesses and heart attacks linked to air pollution. Moreover, 250,000 children a year fall victim to aggravated asthma and other respiratory disorders caused by breathing toxic air—an increase of 11 percent since 1980. Respiratory problems are now the leading cause of children's hospital admissions, according to the National Institutes of Health.

Yet, the deterioration of the nation's air has aroused little attention from the major media outlets. Since 1987, the *New York Times* and the *Washington Post* have printed more than 250 stories on the admittedly serious problems of global warming and acid rain, but have run fewer than two dozen articles on the adverse health impacts of dirty air.

Reprinted from Jeffery St. Clair, "Blowing Smoke," *In These Times*, February 3, 1997, by permission.

So it was no surprise that the Environmental Protection Agency's (EPA) recent announcement of the first new air-pollution standards in more than a decade received only cursory coverage. Of course, this may have been what the Clinton administration was hoping for when it chose to quietly issue its long-awaited new report on air quality on a dead news day: the Friday after Thanksgiving, November 29, 1996, when most Americans were out shopping their way into the next holiday.

> *"The deterioration of the nation's air has aroused little attention from the major media outlets."*

In case you missed it, the EPA recommended a modest tightening of standards on urban smog and fine-particle air pollution, recommendations that EPA Director Carol Browner said were the result "of the most exhaustive scientific review in EPA history." The proposed standards are far from the tight controls hoped for by most clean-air advocates. Nonetheless, early estimates suggest that the new regulations, if adopted, will result in 300 new areas being designated as having chronically unhealthy air, including St. Louis, Louisville, Cleveland, Cincinnati, Des Moines, Indianapolis and New York City. Under the new guidelines, these cities would be required to develop EPA-approved air-quality management plans to reduce emissions or risk losing federal funding for highways and other projects. [Browner signed the new standards on July 16, 1997.] . . .

Smog, of course, is the sepia-colored haze that shrouds Los Angeles and dozens of other cities for much of the summer. It forms in the lower atmosphere where gases such as nitrogen oxides and volatile organic compounds react in warm temperatures. The ingredients for smog are spewed into the air from car exhausts and the smokestacks of oil refineries, power plants and incinerators.

Particulate pollution comes in two varieties: fine and coarse. Coarse particles, more than 2.5 microns in diameter, are typically generated by crushing operations, vehicle travel on unpaved roads and windblown dust. Fine particles are defined as being less than 2.5 microns across, a fraction of the width of a human hair. In contrast to coarse particles, these microscopic particles typically derive from sulfur and nitrogen produced by refineries, smelters, power plants, slash burning and incinerators. On an especially smoggy day, the fine and coarse particles mix together in a kind of airborne chemical cocktail—a single breath can take in millions of these ultra-fine particles. They penetrate deep into the lungs, lodge firmly in vital tissues and stay there, accumulating, for years.

Though the Clean Air Act of 1970 requires the EPA to re-evaluate its air standards every five years, the agency's last assessment of air pollutants was in 1987. During the Bush era, Dan Quayle's off-the-shelf anti-regulation operation, the White House Council on Competitiveness, quashed new reviews of smog and soot regulations—over the rather meek protestations of then EPA head William K. Reilly—at the request of the American Petroleum Institute.

Clinton's EPA, the leadership of which was hand-picked by that green guru,

Al Gore, has hardly been in a rush to correct the damage. In fact, the administration only reassessed the old standards and developed new regulations when it was forced to by a federal court order stemming from a 1992 lawsuit filed by the American Lung Association. The Clinton administration then moved glacially to develop the new regulations, waiting until the last possible moment to release its draft proposal.

This was not always the plan. Some administration officials had wanted to establish new air standards on smog and toxic soot as early as the fall of 1993. Former White House aide Harold Ickes, for example, pointed out that in contrast to earlier air-pollution battles over acid rain and global warming, attacking smog and particulates would, with the right spin, be a publicly popular move because of the tremendous health benefits. Other officials suggested that new air standards could be linked to health care reform.

But this strategy was quickly crushed by a powerful coalition of cabinet members led by the late Ron Brown and former Treasury Secretary Lloyd Bentsen, the oil man from Texas. Bentsen and Brown maintained that such a move would alienate some of the administration's biggest supporters in the business community, including Arco and Southern California Edison. Since Clinton's election in 1992, Arco has sluiced more than $100,000 a year in soft money to the Democratic Party. Its refineries in California would be a major target of the new air regulations.

> *"The regulations . . . only went half as far in restricting smog and fine-particle pollution as the EPA's own scientists had recommended."*

The smokestack industries had good reason to be nervous about the possibility of new clean-air standards. For nearly 20 years, the EPA has focused on coarse particles, such as road dust, fly ash, cement kiln dust and other construction-related pollution. A shift in emphasis to fine particles would affect emissions from power plants, incinerators and other major industries that have until now largely escaped scrutiny.

## The Media's Industrial Bias

What press coverage there was in November 1996 tended to focus on the economic costs of the proposed new restrictions on smog, which the *Chicago Tribune,* quoting an American Petroleum Institute study, said might cost the city of Chicago as much as $7 billion. Other newspapers, such as the *Portland Oregonian* and the *Indianapolis Star,* decried the oppressive nature of the new regulations and their chilling effect on local industries, such as pulp mills, coal-fired power plants and cement kilns. The *Los Angeles Times,* long an implacable critic of federal clean-air laws, quoted city air-management officials as saying the new standards were "simply unattainable."

This news slant was a PR triumph for the energy, incineration and automobile industries, because the real story on the new air regulations is that they aren't nearly as restrictive as most in the environmental and scientific communities had hoped. Indeed, the regulations announced by Browner only went half as far in restricting smog and fine-particle pollution as the EPA's own scientists had recommended.

The energy and manufacturing lobbies worked hard to earn such sympathetic and alarmist press coverage. For the previous six months, they had been furiously feeding the nation's environmental reporters white papers on how the impending regulations were onerous, impractical and unjustified. Industry press releases, using numbers concocted by the creative accountants at PR houses such as Burson-Marsteller and Hill & Knowlton, exclaimed that the cost of further clamping down on smog would tally in the tens of billions and yield negligible environmental benefits. Corporate executives at Chevron, Pacific Gas and Electric, and WMX Technologies warned their shareholders and customers that tougher standards could mean only two things: higher prices and layoffs.

## Attacks on the Regulations

Then, in the wake of the report's release, a squadron of industry lobbyists was deployed to discredit the mounting health data. At the same time, they stressed that the utilities, steel plants and refineries had done all they could do, and that once again the regulatory ax should fall on the motorist and the dry cleaner down the block.

One of the first to attack the science on which the regulations were based was Paul Bailey, health and environmental affairs director at the American Petroleum Institute, the $50-million-a-year trade association for domestic oil companies. Bailey, sounding as if he had just emerged from a seminar at the Tobacco Institute, claimed that urban smog was no big deal and people seemed "actually to adapt to it." On the subject of fine particulates, Bailey asserted that the "science is simply not good enough" to reach any regulatory conclusions.

Using the new air regulations as a pretext, Rep. Tom DeLay of Texas, the former exterminator elevated to House whip, strutted in front of the TV cameras to announce plans to overhaul the Clean Air Act and to strip the EPA of its authority to implement new regulations. DeLay's south Houston district, one of the more toxic in the nation, is packed with oil refineries, incinerators and other industries that would be hit by the new air standards.

> *"An effective clean-air program depends on a rational and progressive energy policy."*

Of course, not all of the industry's lobbying activities were quite so confrontational. In order to spread its message to Democratic politicians and Clinton staffers, the American Petroleum Institute secured the services of Beckel and Cowan, a lobbying outfit run by Democratic Party operative—and frequent *Crossfire* co-host—Bob Beckel. The incineration lobby also placed its interests in extremely capable Democratic hands. Browning-Ferris Industries, the nation's second-largest waste disposal company, hired Peter S. Knight, a lawyer and lobbyist with the powerhouse D.C. firm of Wunder, Diefenderfer, Cannon and Thelan. Knight was an aide to Al Gore during his years in the U.S. Senate and served as chair of the Clinton-Gore re-election campaign.

## A Weakened Policy

These investments paid off. Since the announcement of the proposed air standards, Browner has gone out of her way to assure industry leaders that the new regulations will not be imposed in anything resembling a draconian manner. Instead, she promises to "work state by state, city by city, and industry by industry to develop a common-sense and cost-effective option for implementation of the standards." The new regulations, she has told industry and municipal leaders, are still works-in-progress—meaning there's a good chance that they will be whittled down even further.

Under the most optimistic scenario, smog-plagued cities won't have to submit plans to meet the new standards until 2002, and even then they won't be held accountable for meeting such standards for another decade. Such solicitousness toward the oil lobby undermines the integrity of the Clean Air Act, which is one of the few environmental laws that instructs the EPA to develop regulations that protect the public health without regard to economic costs.

Worst of all, the new air regulations, weak as they are, exist in a political vacuum. An effective clean-air program depends on a rational and progressive energy policy. Such a policy would actively encourage conservation and the use of clean fuels, more efficient automobiles, new investments in public transportation and a wholesale commitment to renewable and clean power sources, such as wind, geothermal and solar. Instead, the opposite is occurring. The budgets for all of these programs are being savagely slashed and the regulatory framework under which they exist—and, theoretically, at least, could be resurrected—is being steadily dismantled.

So industry seems to have won by its usual tactics of whining, influence-peddling and delay. But what must not be forgotten is that further delay has consequences. Nearly 200 Americans die each day from filthy air. Between 1997 and 2002, when the smog and soot plans are supposed to be in place, air pollution will cost 360,000 lives. As always, the heaviest burden will be borne by the most vulnerable: the elderly, the infirm and inner-city children, who have already been savaged by the welfare bill and pernicious cuts in education, housing and health services.

# The Federal Government's Air-Quality Standards Are Too Stringent

## by James M. Inhofe

**About the author:** *James M. Inhofe is a Republican senator from Oklahoma and chairman of the Senate Environment and Public Works subcommittee on Clean Air, Wetlands, Private Property, and Nuclear Safety.*

The Environmental Protection Agency's, or EPA's, proposed changes to the nation's air-quality standards are the most expansive and expensive environmental regulations in history. Based on inconclusive and unsubstantiated science, the rules are premature at best. If adopted, they will result in tremendous costs—up to tens of billions of dollars annually—imposed on state and local governments as well as on individuals, businesses and communities. The likely result: More states threatened with loss of federal highway funds, more businesses forced to absorb costs of expensive new technologies and more drivers facing mandatory emission-control requirements.

## More Studies Are Necessary

The proposed new standards for ozone (smog) and particulate matter (small airborne particles) supposedly are based on conclusive scientific studies, but the only real consensus among scientists is that more study is needed. For 25 years, air quality steadily has been improving around the country. Yet, now we are told drastic new steps are needed to protect "public health." In fact, the suggested marginal health benefits of the new standards are open to serious question and debate.

In addition, the arbitrary and rushed procedures by which these standards have been promulgated undermine public confidence in government and feed a growing public mistrust of all environmental laws. The appropriate course of action would be to hold off on these new standards and allow more public comment and detailed scientific study.

James M. Inhofe, "Q: Are More Rigorous Clean-air Standards Really Necessary? No: Federal Regulators Are Set to Choke Local Economies for the Sake of Marginal Health Benefits," *Insight*, April 28, 1997. Reprinted with permission from *Insight*. Copyright ©1997 News World Communications, Inc. All rights reserved.

As chairman of the Senate Clean Air subcommittee, I have paid close attention to the EPA's proposals and have sought to involve our members in an active oversight role. We have conducted public hearings on the scientific issues behind the rules and on the effects anticipated by affected states and local governments. This process has raised public awareness of a number of key issues and concerns. What has become clear is that many unanswered

> *"For 25 years, air quality steadily has been improving around the country."*

questions remain—from both Democrats and Republicans in Congress—with regard to the regulations and the procedures by which they were developed.

The Clean Air Act requires the administrator of the EPA to examine the current air-quality standards every five years and to assure that, based on the latest scientific evidence, human health is being protected with an adequate margin of safety. With regard to the particulate-matter, or PM, standard, the EPA is under a court-ordered deadline to issue a decision by July 1997. [EPA director Carol Browner signed the new standard on July 16, 1997.] This means that the administrator must decide whether to retain, tighten or relax the current standard. It is not a mandate to change the PM standard.

Clearly, court-ordered deadlines are not conducive to formulating deliberate and thoughtful public policy. In 1994, the EPA argued to the court that it would take at least four years to make a scientifically sound and defensible decision on particulates, yet only two years later the agency proposed these regulations. And, it seems curious that the EPA would seek increased funding in its 1998 budget to study the "uncertainties" of particulate-matter health effects while claiming that the science on this issue already is definitive.

Under the Clean Air Act, scientific assumptions are reviewed by a panel of experts known as the Clean Air Scientific Advisory Committee, or CASAC, which then reports its findings to the administrator. The last review of the particulate-matter standard in 1987 took five years. This time, the review only lasted two years and at every step the CASAC panel noted that it needed more time.

Since the recommendations of the CASAC panel have been cited by those justifying the proposals, our subcommittee received testimony from the chairman and immediate past chairman of the panel during our science hearing. We learned that CASAC's official decision was that the administrator had sufficient information to make a policy decision, but that there was less than full consensus about the science. Individual members of the panel then were polled for their personal opinion about the agency's recommended range for a new standard.

It widely has been reported that 19 out of 21 members endorsed a new PM standard. But this is misleading, since the recommendations ranged from the very strict PM standard adopted by the EPA to a much less stringent standard. In fact, eight members recommended against the manner in which EPA sets the more stringent standard.

In addition to a better understanding of the CASAC decisions, our science hearing also uncovered a number of scientific uncertainties. The research behind the PM proposals are epidemiological studies, which only show a statistical (not a cause and effect) relationship between PM and health effects such as respiratory illness. For example, it is unknown which of the PM particles might be causing the health problems, which size particles are a concern (the proposal regulates the fine particulate size) or even if particulates actually are to blame for the health complaints. Some studies have shown that other pollutants are the cause of the problems or that other variables such as humidity or heat (which were not controlled for in the EPA studies) might pose the real threat to public health. In addition, the agency did not consider indoor air even though the affected people—those displaying the defining health effects—largely are hospital patients who primarily are breathing indoor, not outdoor, air.

## No Increase in Health Benefits

Are higher standards needed to control smog, which directly is related to high ozone levels? It may be argued that the current ozone standard, which many metropolitan areas have yet to attain, has obliged many communities to seek cleaner air quality. But the proposed change in the standard would triple the number of nonattainment areas with little to no increase in the health benefits. Such areas could lose federal highway funds and become subject to onerous mandatory federal controls. Cities that cannot attain the current standard would have no chance of ever meeting a tighter standard.

While some have argued that industry will develop new technologies to reach attainment, this is dubious at best. Many urban areas of the country have been in nonattainment with the current standard since the Clean Air Act first was written, and scientists have been working for 20 years to develop new technologies. No evidence suggests a further tightening of the standard somehow will cause a miraculous technological breakthrough.

While the witnesses at our science hearing disagreed on many points, there was near-unanimous agreement that at least five more years of research and monitoring were necessary to answer some of the major problems and concerns for particulate matter.

State and local government officials have indicated that the effects of being thrown into noncompliance with the new standards would be devastating to their communities and that they are not convinced of the

> *"Cities that cannot attain the current standard would have no chance of ever meeting a tighter standard."*

magnitude of the claimed health benefits. They also are concerned that scarce resources unwisely will be diverted away from more pressing environmental priorities such as safe drinking wafer and the cleanup of toxic-waste sites. The new standards will make it more difficult, if not impossible, for new industrial, manufacturing and service facilities to become established and for existing ones to expand. It doesn't take a rocket scientist to see that local economies will stagnate and jobs will be lost.

Throwing more areas into nonattainment with the new standards only will decrease the incentives for compliance. Governors testified that the stigma of being designated a nonattainment area leads to the inability to attract new businesses or retain existing ones. The unseen but real effect will be more foreign competition for facility locations and jobs.

## The EPA's Dangerous Tactics

Equally disturbing is the process the EPA has used to propose the new standards. Not only has the EPA violated the Unfunded Mandates Law and the Small Business Regulatory Enforcement Fairness Act, it has moved with arbitrary haste and little regard for dissenting views. Indeed, it apparently even has suppressed information contradicting its proposals. As a result, it has acted in a manner that undermines public consensus in what it is trying to do and fosters cynicism about politicization in the federal bureaucracy.

Clearly, the new standards will result in the imposition of new federal mandates. States will be required to implement monitoring programs, develop implementation plans and enact measures to ensure compliance, all of which will cost substantial sums of money. Individual citizens and small businesses directly will be affected through substantial increased costs or unspecified lifestyle changes.

We have heard allegations by other government agencies that the EPA has sought to suppress dissenting views from other agencies within the government. While the EPA has estimated the costs at a maximum of $6 billion annually for ozone alone, the president's own Council of Economic Advisors has estimated the costs to be at least $60 billion per year. Materials prepared by the President's Office of Management and Budget were censored by EPA officials because they "could undermine the Agency's efforts."

In addition, the data from the major research study the EPA depended upon have not been released to the public. The studies in question were conducted in a cooperative effort between researchers at Harvard University and the EPA. The raw data have been requested through the Freedom of Information Act and have not been made available to independent scientists. While contractual requirements may prevent the EPA from sharing the data, it owes the American public access to information used in setting major national policies such as these regulations. Surely, Americans have a right to know how the regulations under which they must live are developed.

No one is advocating a retreat from protecting our nation's air, but so far the agency's proposals raise more questions than have been answered. The current laws and regulations are working and our air is getting cleaner. There simply is no urgency for us to act in a precipitous and unwise manner.

# Electric Cars Reduce Air Pollution

**by Robert W. Benson**

**About the author:** *Robert W. Benson is a professor at Loyola Law School in Los Angeles.*

I have been commuting to downtown Los Angeles since 1991 in my electric car, getting the thumbs-up from other drivers, happy that my converted Volkswagen Karmann Ghia (license plate SMOGLSS) emits not an iota of tailpipe pollution. So, I was aghast to read a report from Carnegie-Mellon University asserting that electric cars may make the environment dirtier. "If this is true," I told my wife, "I'll get rid of the car tomorrow."

## Cleaning Up an Environmental Mess

After all, I bought SMOGLSS only because I felt that as a Los Angeles driver I had contributed more than my lifetime share to dirtying the air. As a professor of environmental law, I thought the time had come for me to practice what I teach. I could not quite summon the virtue of my few colleagues who take the bus, or the campus priest who pedals 25 miles to work on his bicycle, but I figured that the electric car would help get me back in line with the kindergarten rule that you clean up your own messes.

I wanted to be sure, so I delved into research data. I found that my car cuts air pollution (volatile organic compounds, nitrogen oxides, and carbon monoxide) by 97 percent compared with a gas car. I wondered if I was merely transferring the pollution from the city to the distant power plant generating the electricity. No. Only 3 percent as much is emitted at the electric plant to charge my car as is emitted from a gas car.

Cutting air pollution this much means that the electric car is causing 97 percent less cost to the public in form of bad health, reduced property values, crop and forest losses, and damage to buildings. Economists have put a price tag of nearly $15 billion a year on those costs of automobile smog in Los Angeles. That works out to more than $2,100 in costs to other people's health and property caused ev-

Reprinted from Robert W. Benson, "Lead Pollution from Electric Cars? Look Closer at the Facts," *Christian Science Monitor*, June 9, 1995, by permission of the author.

ery year by the average gasoline car doing my 60 mile round-trip commute.

Finally, I determined that the electric car cuts carbon dioxide by 63 percent, fulfilling our nation's pledge under the International Climate Change Convention to reduce greenhouse gases and the risk of global warming.

## A False Report

Now comes the report in *Science* magazine by three professors at Carnegie-Mellon University in Pittsburgh. "A 1998 model electric car," declare the professors, "is estimated to release 60 times more lead per kilometer of use relative to a comparable car burning leaded gasoline." Not that electric cars emit lead at the tailpipe, they say, but lead is emitted where their batteries are made. Egads, I thought, I'm a lead polluter!

Fortunately, the *Science* study is fast going up in smoke. In its last footnote, the authors acknowledge grants from the National Science Foundation (NSF) and thank the Green Design Consortium of the Carnegie-Mellon University Engineering Design Research Center. Nowhere in the article, but buried in the authors' NSF Grant summary available on the Internet, is this astonishing statement: "The Ford Motor Company will work with us in transferring the research results. . . ." Moreover, the university research center's directory of industry affiliates lists, among others, British Petroleum America, Exxon Research & Engineering, General Motors Delco Chassis, General Motors Packard Electric, Mobil R & D, and Shell Development.

The auto and oil companies, of course, have lobbied to scuttle environmental laws in California, New York, and Massachusetts requiring electric cars to be produced by 1998. [California and Massachusetts have waived that requirement.]

A Carnegie-Mellon publication states that, "Green Design Consortium Membership is open to industrial partners interested in participating and guiding consortium projects. Annual membership is $20,000 for

> *"The electric car is causing 97 percent less cost to the public in form of bad health."*

new members, and $10,000 for pre-existing members of the Engineering Design Research Center. . . . Membership benefits include: The opportunity to provide input on research direction and suggest specific research programs. Access to: Carnegie-Mellon University laboratories and researchers. . . . Two membership meetings per year . . . to ensure feedback of projects . . . and guide project development."

## Carnegie-Mellon's Bad Science

I called one of the study's authors, Professor Francis C. McMichael, who acknowledged oil and auto company membership in the research center. "It is just the nature of research at private universities these days to seek corporate sponsorship," he said. He emphasized that no specific oil or auto company funding was earmarked for this study. Dr. McMichael did not seem bothered by the fact that for $10,000 or $20,000 a year his "industrial partners" can "provide input on research direction" and "guide project development." Thus is the notion of scientific integrity lost. Thus does Carnegie-Mellon risk slipping into science for sale, and then into bad science.

And the Carnegie-Mellon study is bad science. "Sloppy" and "misleading" are the words used by Daniel Sperling, director of the Institute for Transportation Studies at the University of California, Davis. There are outright mistakes, like confusing kilograms for pounds and miscalculating the weight of GM's current battery pack. The "60 times more lead per kilometer" assertion is an Alice-in-Wonderland figure based on at least five major methodological errors.

As with the cold-fusion claim several years ago, the researchers have offered no evidence to support their basic thesis—that lead pollution from making batteries for electric vehicles is more damaging to health and the environment than the lead formerly contained in gasoline, and also more damaging than the smog caused by gas cars.

## A Failure to Calculate Costs

To prove this, they will have to show that lead is escaping from the three primary smelters, 23 secondary smelters, and manufacturing plants in enormous quantities and is affecting the health of millions. They will have to show that health costs of the escaping lead exceed all health and environmental costs from old leaded gasoline and from smog. The Environmental Protection Agency (EPA) says virtually no lead is escaping from these sites. The Carnegie-Mellon scholars guess that some lead is escaping, based on old data from years before EPA controls were in effect. But they never bother to calculate the health costs of the claimed escape and never compare them with the health and environmental costs of leaded gas and smog.

Their thesis is never proved, and their paper, lacking the logical and intellectual integrity of science, deserves a thumbs-down.

# Electric Cars Are Costly and May Not Reduce Pollution

## by Lawrence W. Reed

**About the author:** *Lawrence W. Reed is an economist and author. He is the president of the Mackinac Center for Public Policy, a free-market educational and research organization based in Michigan.*

Should government enact laws to subsidize "alternative" fuels so that people will use less gasoline? Should government mandate the use of electric cars or other vehicles that don't use gasoline at all?

Even without expert knowledge of the issues involved, anyone who values liberty will be inclined to answer both questions in the negative. Subsidies are a forcible transfer of wealth from those who have earned it to those who didn't. Likewise, mandates are edicts that carry penalties for noncompliance. Use of gasoline for transportation purposes would not seem to be the sort of offense that anyone ought to be behind bars for. Both subsidies and mandates are government's way of declaring, "We're smarter than the market, so we're going to have to force the market to change." How many times have we heard that one—and later lamented the results?

## Misguided Subsidies

As it turns out, the skeptics are right once again. Subsidies and mandates on behalf of alternative fuels are yet another public folly, motivated perhaps by good intentions but fraught with inherent contradictions. Not only do they whittle away at personal liberty, they flout economics and science as well.

Subsidies and mandates for alternative fuels are being discussed in many state legislatures now. A law passed by Congress in 1992 mandated that 75 percent of the half-million vehicles the federal government maintains be fueled by something other than gasoline by 1999; it also required state governments to

Reprinted, by permission, from Lawrence W. Reed, "The Electric Car Seduction," *The Freeman*, November 1996.

start changing the composition of their own fleets to favor more alternative-fuel vehicles (AFVs).

A California law would have forced automakers to begin selling thousands of electric cars to the general public there in 1998, even though in 1995 there were fewer than 1,000 such vehicles on the road in the entire country. By 2003, electric cars were to comprise 10 percent of all cars sold in California—at least 200,000 vehicles. In the wake of withering criticism from scientists, engineers, and economists, the state has backed off somewhat, and the 10 percent mandate will now take effect in 2005. General Motors, Chrysler, and Ford, as well as the big automakers of Japan, will face fines of $5,000 for each car under the required threshold.

> *"The plants that make the power to recharge [electric cars] may put more pollutants in the air than gasoline-powered vehicles."*

Electricity from batteries is one of several alternative fuels that some legislators want to force-feed the marketplace. Others include methanol, ethanol, and other alcohols, hydrogen, compressed natural gas, coal-derived liquid fuels, and liquefied petroleum gas. All of them are championed as fuels that will reduce oil imports, give the economy a boost, and cut pollution.

## Government Should Not Be Involved

Skepticism is countered with a line of reasoning that goes something like this: Good intentions plus the force of law equals positive results. If the market won't do it voluntarily, something must be wrong with the market. Government says AFVs are a good thing and must order the marketplace to comply. We're all better off as a result of such wise intervention because governments are especially good at foreseeing both the future and the policies needed to make it better. What could possibly be wrong with this picture?

The truth is, skepticism about a product is warranted any time it takes special favors or penalties against the competition to gain acceptance for that product. The questions people need to ask are these: *If alternative fuels are all that they are cracked up to be, why do politicians have to get involved? Why can't AFVs succeed on their own?*

The case for electric cars, it turns out, is more hype than substance. They are extremely expensive—costing between $28,000 and $48,000 to make. The batteries, which start at $1,500, must be recharged for at least five hours after driving less than 150 miles. And while the cars themselves are nonpolluting, the General Accounting Office in Washington says the plants that make the power to recharge them may put more pollutants in the air than gasoline-powered vehicles.

Meanwhile, the cleaner fuels used in today's more efficient cars emit a small fraction of the pollution cars did 30 years ago. Even without any conversion to AFVs, the new cars coming off assembly lines by the end of this decade will simply not pose a pollution problem worth worrying about. Analysts at the Reason Foundation in California argue that alternative-fuel mandates would price new cars out of the reach of many Americans, who would then keep their older, more polluting cars longer.

## Less Costly Solutions Exist

Furthermore, according to Charles Oliver of *Investor's Business Daily,* a University of Denver chemist found that the dirtiest 10 percent of all cars produce half of all airborne emissions. Relatively inexpensive, low-tech, voluntary approaches could take care of most of that without the intrusive, high-cost problems of electric-car mandates.

The other alternative fuels in use or on the drawing boards share at least some of the same problems. They require huge subsidies to hide their real costs and make them seem affordable, or they have limited application, or they cause substantial consumer inconvenience, or they generate unintended, harmful side-effects. Time and technology may work the bugs out and make them feasible, but that process will be stymied or misdirected if the nearsighted bias of politicians overrules the marketplace.

When it comes to the government planning of the nation's technological future, we should learn from past mistakes. "Washington is no more capable of forcing industry to develop viable and environmentally-friendly technologies," says James Sheehan of the Competitive Enterprise Institute, "than it is capable of centrally planning the economy." Federal technology programs like the breeder nuclear reactor, the supercollider, and the synthetic fuels program wasted billions of dollars and produced few, if any, tangible benefits.

Patrick Bedard put it well in a December 1993 article in *Car and Driver* magazine entitled "Why Alternative Fuels Make No Sense." He stated, "The promises made for the alternatives to gasoline are very seductive. And you know how seductions turn out."

If politicians continue to micromanage the marketplace to artificially benefit alternative fuels, you may want to reach for both your wallet and your oxygen mask.

# The Clean Water Act Has Improved America's Rivers and Lakes

## by Paul Schneider

**About the author:** *Paul Schneider is an environmental writer and the author of* The Adirondacks: A History of America's First Wilderness.

The Clean Water Act is an immense piece of legislation, with more than 500 sections, so summarizing it is difficult. Suffice it to say that before it was passed there were no enforceable national standards for industrial or sewage discharge into surface waters; now all such "point sources" of pollution require state- or Environmental Protection Agency (EPA)-issued permits. The act also established a national policy on the protection of wetlands, the crucial foundation to healthy surface-water ecosystems, which were filled in or drained at a rate of about half a million acres a year between 1950 and 1970.

Just as important as the regulations was that for two decades Congress was willing to spend public money to carry out the act's mandate. Between 1972 and 1989 the EPA spent roughly $54 billion and the states were required to spend another $128 billion on new or upgraded municipal sewage-treatment facilities; in more recent years another $19 billion has been invested in revolving loan funds intended to provide permanent sources of funding for municipal-waste-treatment improvements.

## Successes and Failures

The result, according to the EPA, is that even though the amount of treated sewage increased 30 percent between 1970 and 1985, there was a 46 percent reduction in the amount of organic waste released into surface waters. Controls established under the Clean Water Act have prevented the dumping of about 1 billion pounds a year of toxic pollutants. According to the EPA, more than 90 percent of the pollution coming from point sources like factories and municipal

Reprinted, by permission, from Paul Schneider, "Clear Progress," *Audubon*, September/October 1997.

waste-treatment plants has been eliminated.

These are stunning achievements, befitting an act that was passed in a post-Apollo-moon-mission era of optimism about the ability of the government to make technical promises and keep them. Yet despite all the progress, in its most recent biennial report to Congress, in 1994, the EPA found that "about forty percent of the nation's surveyed rivers, lakes, and estuaries are too polluted for basic uses." That's far short of the Clean Water Act's original promise that "wherever attainable" rivers and lakes would be safe for swimming and fishing by 1983. . . .

It's no real mystery why the noble effort to "restore and maintain" the nation's waters has stalled. Chief among the reasons is the Clean Water Act's continuing inability to limit urban, suburban, and agricultural runoff, what is known as nonpoint-source pollution. Every time it rains or a field is irrigated, sediment, fertilizers, and pesticides trickle into streams, rivers, and lakes. Fertilizers load the water with nitrates and phosphates, causing boom-and-bust cycles of algae growth and decay that consume the available oxygen in the water.

## The Causes of Contamination

But despite the fact that agriculture is the leading source of water pollution in the United States, responsible for the contamination in 60 percent of the nation's degraded rivers and in half its impaired lakes, the industry is exempt from the permitting process that regulates point sources of water pollution. Farmers are also largely exempt from the wetlands provisions of the act.

Fixing this isn't simply a matter of U.S. farms "going organic," though that would certainly help. Perfectly natural cow manure loads streams with nutrients and fecal-coliform bacteria if animals are not fenced out of every little stream and brook or if feedlots are not properly engineered and managed to contain runoff. Even plain old dirt does more than turn the water brown: It blocks light needed by aquatic plants and interferes with the reproduction of many fish and invertebrates. Sediment, in fact, is the leading cause of water-quality deterioration.

And farms are far from the only source of poisoned runoff. Most people wouldn't drink the water draining off a mall parking lot or a 21-pump gas plaza, but chances are quite good it flows into a stream or river that the Clean Water Act charges the EPA and the states with someday restoring to drinking-level quality. Scenic roads along rivers contribute salt and sand. The tens of thousands of miles of dirt

> *"For two decades Congress was willing to spend public money to carry out the [Clean Water Act's] mandate."*

roads built by the U.S. Forest Service and the logging companies it serves are notorious sources of sediment in otherwise pristine waters. Construction sites erode at a rate more than 1,000 times that of forestland, and all over the West, abandoned mines leach heavy metals and other pollutants into water supplies.

Finally, there's air pollution, the ultimate nonpoint source and the leading cause of continuing contamination in the Great Lakes, where airborne toxins fall into the water and fish-consumption advisories are currently in effect for 97 percent of the shoreline.

For fairly straightforward reasons, foul runoff is much more difficult to control than point-source pollution: There's no gushing pipe that regulators can point to, no corporate board of directors or town sewage authority to threaten with legal action, and often no way to conclusively prove where a given pollutant originated. But even if it were possible or desirable to

> *"The problem isn't what we do to the water, it's what we do to the land."*

identify and regulate every farm, golf course, and backyard from which pollutants leach or erode into the nation's waters, progress would likely be painfully slow. The implications of controlling poisoned runoff are enormous, simply because the problem isn't what we do to the water, it's what we do to the land.

So behind the story of a generation of Americans who, armed with a powerful new law, are reclaiming their inherited wealth of rivers and lakes lies another, more complicated tale—one of money, politics, old habits, and new ways. All over America, as the following three case studies show, there is much to celebrate after a quarter-century of effort. And there is much still to do.

## The French Broad River

Bill Allen remembers the bad old days on the French Broad River. The way the Swannanoa, a major tributary that enters the river in his home city of Asheville, North Carolina, used to run different colors—red, green, yellow, blue—depending on what color blankets were being produced at the mill. The way it pretty much ran black the rest of the time. No vegetation grew along the banks below the big bleaching plant in town, and some days the air around that stretch would burn your eyes. Allen remembers the time when the farm crews over at the Biltmore estate, where his father and grandfather worked and where he grew up, tried to use water from the French Broad to irrigate a crop of corn; the plants withered within a few days. . . .

But industry alone didn't destroy this fast-moving river, which flows out of the Pisgah National Forest in the Smoky Mountains, down through dramatic whitewater gorges and lovely farm valleys, 192 miles across the Tennessee line to Knoxville, where it joins the Holston to form the Tennessee River. Until the Clean Water Act forced the state and local governments to take action, the French Broad was also the primary sewage-treatment facility for all the towns along its banks.

"When I first started looking at this river, back in the mid-seventies, the common knowledge was that you wanted to avoid whole-body contact because the bacteria counts were so high," recalled Richard Maas in his office at the University of North Carolina at Asheville. Maas, an aquatic chemist, was on the board of the local sewer authority from 1989 to 1993. The city's problem then was that even though the quality of the treatment plant had improved dramatically, with federal help, in the first decade after the passage of the Clean Water Act, Asheville was riddled with rotten, leaky sewage pipes. "Just a mile from here they found a line with a hole this big," Maas said, holding his hands apart, "and they dug it up and the entire million and a half gallons of raw sewage just disappeared into this huge underground cavern. It was in the stream within minutes."

## Asheville's Efforts

So Asheville embarked on a breakneck infrastructure upgrade that has already had a remarkable impact on water quality. Local officials also used the power of the Clean Water Act to go after toxic releasers and force them to pretreat their wastes. Most companies, such as the blanket factory, were responsible corporate citizens and complied, but the owner of an electroplating firm went to jail after a six-month FBI investigation proved he was surreptitiously pumping chromium-laden waste down a bathroom drain.

Maas is quick to note that the French Broad is far from running truly clean. From the earliest days of nonnative contact with the river, when burly drovers herded swine and turkeys up the valley to feed the slave-and-tobacco economy on the other side of the mountains, local agricultural practices have added hefty loads of sediment and manure to the mix. Matching federal and state funds are beginning to encourage some large farmers to build manure-settling lagoons and fence animals away from streams, but progress is slow among the many smaller farmers. Just as worrisome is the explosive population growth in the region, with its attendant subdivisions, malls, golf courses, and increased sewage. "Land use equals water quality," Maas likes to tell his students.

Nor is point-source pollution entirely gone from the watershed; the small-time metal plater went to jail, but big players such as Accousta and the Champion paper company have permits from the EPA that allow them to "release"—not dump, of course—pollutants into the river. They've spent millions of dollars on improvements and are doing a far better job than they were before the Clean Water Act, but you need only go to the giant Champion plant on a major French Broad tributary called the Pigeon River to see the problems that remain. There's a public school just upstream from the factory, and you can stand on the riverbank and see down two or three feet to good-size trout hanging in the current. Just below the plant, on the other hand, the river is a murky brown mess. Locals say that little other than catfish and suckers lives in the Pigeon below the plant, though they quickly add that that's more than lived there 15 years ago. Downstream in

*"Despite the lingering problems, those who remember the old French Broad are optimistic."*

Knoxville, which gets its drinking water from the French Broad, concerned citizens have formed a Dead Pigeon River Society to try to get the river cleaned up.

Despite the lingering problems, those who remember the old French Broad are optimistic. . . .

## The Development of the Boise River

A generation ago, most people in Boise barely took note of the Boise River, which runs just south of downtown; it was an urban-planning consultant from California who suggested in the 1960s that the city consider creating the greenbelt along the river. . . .

To its credit, Boise began the process of reclaiming its river even before the passage of the Clean Water Act; the city government decided to follow the consultant's advice and direct development away from the riverfront. But the regulatory muscle of the federal law—and more important, the matching dollars for a new municipal waste-treatment facility—were crucial to raising the water quality to a level where people could take full advantage of the new waterfront parks without fearing for their health.

"Boise succeeded in capturing and preserving most, maybe all, of the nonmonetary values associated with the river ecosystem," says David Eberle, a local economist who thinks the city's river-development plans may be a model for other places. "And the payoff has been, among other things, significantly higher real estate values."

> *"To its credit, Boise began the process of reclaiming its river even before the passage of the Clean Water Act."*

Eberle's enthusiasm is not shared by everyone in the local conservation community. Some point to the daily, almost visible increase in suburban sprawl as a sign of ominous nonpoint water-quality problems to come. But the cutely named subdivisions and "private neighborhoods" creeping out from town in both directions along the greenbelt may be better for the river than the farms they replace. After flowing through the city, the river meanders west through a rich agricultural valley. There are dairy and beef operations where you can see cattle wallowing right in the tributaries. All over the valley are irrigation ditches where water from the river is diverted and allowed to flow across fields, watering the crops but also picking up the usual brew of sediment, manure, bacteria, and chemicals. By the time the Boise gets to its juncture with the Snake River, at Idaho's border with Oregon, it's officially listed as "water quality limited" in nearly every category, including nutrients, sediment, temperature, fecal-coliform bacteria, and dissolved oxygen.

## A Debate Over Requirements

In response, the state Division of Environmental Quality is developing a watershed-wide plan for decreasing the total maximum daily load of pollution in the Boise River. Officials are attempting to include all the relevant "stake-holders" in the effort, as the Clean Water Act requires. And not only for the Boise. In a recent lawsuit brought by the Idaho Conservation League, a federal court ruled that the EPA had failed to enforce the Clean Water Act in Idaho; the state must now rapidly prepare watershed plans not just for the 36 stretches of river it had hoped to clean up but for more than 900 other degraded waterways.

Such requirements are viewed by many in Idaho as an intrusion of federal power. "It is the heavy hand of government," said Mayor Coles. Never mind that the green pastures and verdant mint and potato fields of the "treasure valley" exist mainly because of the largess of the federal government. Twelve miles upstream from town, the Boise River comes out of a pipe at the base of the first of three enormous irrigation and flood-control dams built with federal money between 1915 and 1955. Wherever you look there's a straight line of color change where the subsidized irrigation stops and the old sagebrush of the high desert begins. Like many other rivers in the west, the Boise is "fully allocated," meaning that every drop that comes out of the pipe at Lucky Peak Dam is guaranteed by right to some person, corporation, or municipal entity. As good as the fishing is in the Boise, it would be much better if the trout could afford to buy a few more water rights.

## Trouble Along the Mississippi

At the upper end of the Mississippi River is a small, limpid Minnesota lake called Itaska. Drinkable, swimmable, fishable, it is as close to pristine as any major tourist destination in America can likely be. For the next 2,350 miles, almost every success and every failure of the Clean Water Act over an area encompassing 40 percent of the United States ultimately affects this river. . . .

The Upper Mississippi is in such bad shape that some government biologists worry that the amazing diversity and quantity of plant and bird life for which it is internationally known may be in jeopardy.

"We're concerned that we may reach a point of no return, where the system rapidly declines and then fails to recover," said John Duyvejonck, a biologist with the U.S. Fish and Wildlife Service. He cited recent declines in submergent plant species, such as wild celery and small invertebrates such as fingernail clams, both of which are fed on extensively by migratory waterfowl, particularly canvasback ducks. He noted that the cottonwood trees, which provide food and habitat along the river's edge and on the many backwater islands, are apparently not reproducing.

## Flood Control and the Environment

But the most pressing problem on the Upper Mississippi, Duyvejonck and most other scientists agree, is not the quality of the water but the fact that the U.S. Army Corps of Engineer manages the river through a system of locks and dams, levees, and wing dams to maintain a nine-foot navigation channel and, further downstream, to control flooding. Indeed, like so many other bodies of water in the United States, the Upper Mississippi *is* notably cleaner now than it was 25 years ago. Before the Clean Water Act, a 64-mile stretch of river below the Twin Cities was for all intents and purposes dead. Thanks to massive spending at all levels of government, the effluent coming from Minneapolis–St. Paul is now highly treated, and storm runoff no longer results in raw sewage flowing into the river.

"Most of us don't ever really even think about water quality," said Barry Drazkowski as his colleague, Rory Vose, guided a pontoon boat out of the main navigation channel and into a backwater near Wynona, Minnesota. Drazkowski, a river ecologist, used to work for the U.S. Geological Survey (USGS). Now he and Vose direct the Resource Studies Center at St. Mary's University of Minnesota. They think and talk about altering the thousands of structures that the corps has built over the past century and a half, so that instead of slowly silting in and becoming a static series of shallow lakes, the Mississippi could again act like a river. They want a dynamic river system that occasionally rages high enough to create productive new habitats and scour out overmature ones, a river that periodically recedes to allow riparian trees and plants to germinate. Although no one realistically believes that major dams will be removed in the foreseeable future, Drazkowski and others are encouraging the corps to take out or change the shape of some of the smaller structures that inhibit the river's natural flow. They also believe that occasionally opening the floodgates and drawing down the pools could produce environmental benefits that far outweigh any short-term disruption of the shipping industry. . . .

## The Future of America's Rivers

In 1995, only the threat of a veto from President Bill Clinton and a flood of public outrage halted a bill passed by the new Republican House majority that would have gutted the Clean Water Act.

In that groundswell, perhaps, lies evidence of both the greatest achievement of the Clean Water Act and the hope for the future of the nation's rivers and lakes. Citizens have noticed the improvement in the past quarter-century, and they are not interested in returning to a polluters' free-for-all. The Clean Water Network, a consortium of citizen groups concerned with water issues, counts more than 900 environmental, sporting, religious, labor, and other organizations among its members. All over the country, adopt-a-river programs are springing up, and although they're not enough in themselves to finish the job of restoring U.S. rivers, they are enabling citizens to take responsibility for their local waters.

But citizen action cannot entirely replace political leadership. Nor can the federal courts, which have several times intervened in cases where the Clean Water Act is strong but political will is weak. For example, municipal waste remains the number-two source of water pollution in the country, after agriculture, and it will cost an estimated $137 billion in the next 15 years just to prevent backsliding on the progress already made in that area. That's almost as much as has been spent on municipal waste treatment since the passage of the act. Despite the current budget-cutting fervor, however, there's reason to believe Congress may find the requisite spine to act in the public interest; 19 incumbents who voted for the 1995 "dirty water bill" lost their seats two years later to opponents who highlighted their environmental record.

The Clean Water Act has not yet achieved its goal of giving Americans back the wealth of clean rivers and lakes that were taken from them over the past two centuries; but then, neither has the Civil Rights Act eliminated a legacy of racism, or the laws against murder ended violent crime. What the Clean Water Act and the resulting significant strides toward eliminating water pollution have done is permanently change expectations, giving the public the idea that progress toward clean water is possible. Not inevitable, not easy, and not cheap. But possible. And right.

# Additional Government Programs Are Not the Correct Response to Water Pollution

by Alex Annett

**About the author:** *Alex Annett is a research assistant for the Heritage Foundation, a public policy research organization.*

During the 1997 State of the Union address, President Bill Clinton announced a new federal program entitled the American Heritage Rivers Initiative (AHRI), which he intended to support communities in their efforts to restore and protect rivers across the United States. To many, this lofty goal sounds good. But, on closer inspection, the pristine image it paints becomes murky, revealing a program that violates many constitutional and statutory provisions, involves the federal government further in local and state environmental issues, is inefficient and wastes tax dollars, and threatens personal property rights. . . .

## The Details of the Initiative

President Clinton unveiled new details about how he plans to implement his new American Heritage Rivers Initiative when he issued Executive Order 13061 on September 11, 1997. Through executive order, Clinton has established an American Heritage Rivers Interagency Committee to oversee implementation of the initiative. Members of the committee will include the secretaries of the Departments of Agriculture, Commerce, Defense, Energy, Housing and Urban Development, Interior, and Transportation; the attorney general; the administrator of the Environmental Protection Agency; the chairpersons of the Advisory Council on Historic Preservation, the National Endowment for the Arts, and the National Endowment for the Humanities; or designees at the assistant secretary level or their equivalent.

Reprinted from Alex Annett, "Good Politics, Bad Policy: Clinton's American Heritage Rivers Initiative," *Heritage Foundation F.Y.I.*, February 2, 1998, by permission of the Heritage Foundation.

To nominate a river for designation as an American Heritage River, a local community must submit a river nomination packet to the President's Council on Environmental Quality. The packet must include: a description of the river or river area to be considered, its notable resource qualities, a clearly defined vision for protecting the area and a specific plan of action to achieve it, evidence that a range of citizens and organizations in the community support the nomination and plan of action, and evidence that individuals in the community have had an opportunity to discuss and comment on the nomination and plan of action.

The Council on Environmental Quality will select a panel of experts to review the nominations and make recommendations to the President. From these recommendations, the

> *"President [Bill] Clinton is claiming for himself and future Presidents powers that belong to Congress."*

President would select ten rivers or river areas to designate as American Heritage Rivers. These American Heritage Rivers would receive preferential treatment for federal dollars and the support of other federal programs.

On the surface, President Clinton's program looks appealing. Rivers have played a vital role in the country's history, culture, recreation, health, environment, and economy. Finding ways to encourage states and local communities across the country to become involved in improving the water quality of their rivers and revitalizing their waterfronts is commendable. The AHRI, however, will amount to little more than a surface ripple in accomplishing these goals. . . .

## A Violation of the Constitution

Americans love the beauty and resources of their country. They clearly understand that the U.S. Constitution establishes a system of government to protect their individual rights, and that the federal government should be expressly limited in its ability to usurp those rights. They may disagree, at times, about how much power is given each branch of the federal government to settle disputes and to limit personal freedoms, but there is no dispute that the Founding Fathers intentionally and explicitly designed a balance of power to prevent legislative, judicial, or executive arrogance and abuse of power. Americans expect their elected leaders to abide by the separation of powers delineated in the Constitution, and they want the federal judiciary on guard to make sure they do.

Rather than honor these expectations, President Clinton's American Heritage Rivers Initiative violates both the intent and the letter of the U.S. Constitution. It gives the President as well as his executive agencies authorities that clearly and constitutionally belong to the legislative branch of government, and it confiscates the land use and zoning powers of the states.

> The Constitution protects us from our own best intentions: It divides power among sovereigns and among branches of government precisely so that we may resist the temptation to concentrate power in one location as an expedient

solution to the crisis of the day.

—*New York vs. United States*, 112 S.Ct. 2408 (1992)

Under the U.S. system of checks and balances, the legislative branch has the power to create laws and appropriate funding, the executive branch is author- ized to implement and enforce the laws, and the judiciary is given power to in- terpret those laws in disputes. To explain to hesitant colonists why this separa- tion of powers was important, James Madison wrote in *Federalist* No. 47 that the "accumulation of all powers legislative, executive and judiciary in the same hands, whether of one, a few or many, and whether hereditary, self appointed or elective, may justly be pronounced the very definition of tyranny."

## Supreme Court Rulings

The Supreme Court historically has recognized the importance of the separa- tion of powers among the President, Congress, and the judiciary. In the case of *Youngstown Sheet & Tube Co. v. Sawyer*, the Supreme Court was asked to de- cide whether President Harry S Truman (during the Korean War) was acting within his constitutional power when he issued an executive order directing the Secretary of Commerce to take possession of and operate most of the country's steel mills. The government's position was that the president's action was nec- essary to avert a national disaster that inevitably would result from the stoppage of steel production, and that in meeting this grave emergency, the President was acting within the aggregate of his constitutional powers. The Supreme Court found in *Youngstown* that, even with the threat of a national catastrophe, the President's order could not be sustained as an exercise of his authority. In this case, the Supreme Court found no statute that expressly authorized the Presi- dent to take property as President Truman's executive order intended, or any act of Congress from which such authority could be inferred. The Supreme Court concluded that the power to adopt such public policies as those proclaimed by the executive order is beyond question by Congress, and that the Constitution does not subject this law-making power of Congress to the President.

Supreme Court precedent suggests that President Clinton's Executive Order No. 13061 runs contrary to the separation of power provisions of the Constitu- tion. To implement the AHRI, President Clinton is claiming for himself and fu- ture Presidents powers that belong to Congress: specifically, authority over in- terstate commerce, water rights, property rights, and the appropriation of money. Through executive order, Congress would be relegated to a role of try- ing to stop presidential programs from being implemented, rather than creating and approving them based on the will of the people and funding them as au- thorized in the Constitution.

The Property Clause in Article IV of the Constitution states that "Congress shall have power to dispose of and make all needful Rules and Regulations re- specting the territory or other property belonging to the United States." Execu- tive Order No. 13061, however, gives the executive branch control and authority over the country's rivers and their associated resources located on federal lands, a power specifically assigned to Congress. In order for the executive branch to have authority to govern and control these rivers and associated resources, this power must be delegated to it by an act of Congress. Congress has not given the executive branch such authority.

The Tenth Amendment to the Constitution stipulates that the "powers not delegated to the United States [federal government] by the Constitution, nor prohibited by it to the States, are reserved to the States respectively or to the people." Under the Tenth Amendment, then, state and local governments retain the authority to engage in land use planning and local zoning for public health, safety, and welfare. President Clinton's program, however, sets a new precedent by giving federal regulators a greater role in land use planning, local zoning, and other aspects of a river's surroundings, including "characteristics of the natural, economic, agricultural, scenic, historic, cultural, or recreational resources of a river that render it distinctive or unique." The President has no authority under the Constitution to engage in land use planning and local zoning; thus, Executive Order No. 13061 violates the Tenth Amendment.

## A Conflict with Two Environmental Laws

In addition to altering the constitutional separation of powers, the AHRI implementation process outlined in Executive Order No. 13061 also conflicts directly with two significant environmental laws: the National Environmental Policy Act (NEPA) and the Federal Land Management and Policy Act (FLMPA).

The Clinton Administration has cited the National Environmental Policy Act of 1969 as the legal basis for establishing the AHRI. The NEPA is primarily a policy statute mandating that federal government agencies consider the environmental effects of major federal actions. The idea behind the NEPA is that, by requiring federal agencies to consider and gather information about the environmental consequences of proposed actions, the agencies will make wiser environmental decisions. President Clinton states that the NEPA provides a grant of authority to establish the AHRI under authority of Section 101(b) of the NEPA. This section only sets out the broad goal to be achieved by the NEPA, however; it provides no authority for action. The only authorities mandated to the executive branch under the NEPA are to prepare reports; interpret and administer federal policies, regulations, and public laws in accordance with the NEPA; provide information, alternatives, and recommendations; and provide international and national coordination efforts. President Clinton apparently has interpreted these duties to mean that the NEPA also gives the executive branch broad authority to develop programs. Such authority, however, was given specifically to Congress, not the

> *"The President has no authority under the Constitution to engage in land use planning and local zoning."*

President, and Congress has not delegated such powers explicitly to the President. Consequently, citing the NEPA as the legal basis for implementation of the AHRI is questionable.

Even if it can be argued successfully that President Clinton's action is consistent with the purpose of the NEPA, the NEPA, as written, does not trump the requirements of other statutes. And, in the case of the Federal Land Management and Policy Act, the President is expressly restricted in his ability to designate or manage federal lands. Congress enacted the FLMPA in 1976 in order to reestablish its authority over the designation or dedication of federal lands for specified purposes, and to circumscribe the authority of the President and executive branch to manage federal lands.

> *"The protection of personal property in the Constitution is under increasing assault by all levels of government."*

In the FLMPA, Congress declared that "it is the policy of the United States that Congress exercise its constitutional authority to withdraw or otherwise designate or dedicate Federal lands for specified purposes" and delineate the extent to which the executive branch may withdraw lands without legislative action. Congress thus asserted its authority to create, modify, and terminate designations for national parks, national forests, wilderness, Indian reservations, certain defense withdrawals, national wild and scenic rivers, national trails, and other national recreational areas and national seashores.

In fact, Congress has not withdrawn, designated, or dedicated any federal lands for President Clinton's American Heritage Rivers Initiative, nor has it authorized the development of the program by the executive branch. The legislative process for obtaining a favored status designation for federal land and resources is clearly established. Consider, for example, the Wild and Scenic Rivers Act adopted by Congress on October 2, 1968. The act provides for the selection, by Congress, of American rivers that, along with their immediate environments, possess outstandingly remarkable scenic, recreational, geologic, fish and wildlife, historic, cultural, or other similar values. The rivers selected are protected for the benefit and enjoyment of present and future generations. Since 1968, Congress has designated 154 Wild and Scenic Rivers under this act, amounting to 10,814 miles of river. In fact, Congress acted as recently as November 12, 1996, when it designated 11.5 miles of the Lamprey River in New Hampshire and 6.4 miles of the Elkhorn Creek in Oregon, following the designation of 51.4 miles of the Clarion River in Pennsylvania on October 19, 1996, as part of the Wild and Scenic Rivers program. . . . Clearly, when Members of Congress believe there is reason to act, they will act.

If President Clinton wants to see his initiative implemented properly, then he first should propose legislation to Congress and allow Congress to approve or reject the initiative based on the merits of the proposal and the will of the people. Because Congress has not designated or dedicated any federal lands for the AHRI, or authorized the development of the AHRI, the actions of the President in creating and implementing the AHRI through Executive Order No. 13061 violate the FLMPA.

## The AHRI Threatens Property Rights

The protection of personal property in the Constitution is under increasing assault by all levels of government. The right to own and use property free from unreasonable or arbitrary government interference is fundamental to American freedom and the U.S. Constitution. In fact, the Framers of the Constitution considered the protection of property rights so important that they included it in the Third, Fourth, Fifth, and Sixth Amendments. Today, in an era of almost daily documented cases of unreasonable and arbitrary interference by government agencies, it is not surprising that the Clinton Administration does not seem to recognize or agree with the Founders on the importance of individual property rights.

This lack of appreciation for personal property rights is an undercurrent in President Clinton's AHRI. The right of individuals who own property along designated rivers to use their property free from unreasonable and arbitrary government interference is threatened by the AHRI. The Administration has resisted adding a mandatory opt-in provision to allow the property of landowners along designated American Heritage Rivers to be included in a nomination only in cases in which owners have given their

*"Water rights and land-use planning . . . are subject to regulation and control at the levels of state and local elected government."*

written permission. Such a provision would have shown that President Clinton indeed was concerned about the property rights of those Americans whose land is located along designated rivers. The lack of such a provision means property owners have no guarantee that their property rights are protected.

The regulation of wetlands under the Clean Water Act affects hundreds of thousands of acres of property across the United States. Implementing the AHRI will add hundreds of thousands of acres of dry land to the federal government's control in perpetuity. Rather than increase the access of people to federal resources and protect their rights, the AHRI will increase the access of federal bureaucrats to private property across the United States.

## Interfering with State Jurisdictions

The Founders believed that government closest to the people works best. The Tenth Amendment addresses the empowerment of state and local communities to govern. It recognizes that the federal government—as an entity—should have only limited powers, and that its powers should be specifically enumerated. Water rights and land-use planning are not stipulated powers of the federal government; historically they are subject to regulation and control at the levels of state and local elected government. As Chief Justice William Rehnquist has argued, taking the control of water from the legislatures of the various states and territories at the present time would be nothing less than suicidal. If the appropriation and use were not under the provisions of state law, the utmost confusion would prevail.

President Clinton, through his executive order, is attempting to establish and exert federal control over something that clearly is under state jurisdiction. By allowing the intervention of the federal government through federal bureaucrats, known as "river navigators," who are appointed by the President, Executive Order No. 13061 will interject the federal government heavily into the local decision-making process.

The Clinton Administration claims that river navigators will not interfere in the local planning and zoning process, yet it resists incorporating a provision to prohibit them and all other federal employees involved with the initiative from intervening in local zoning and other decisions affecting private property and water rights. Such a provision would ensure that the states and local communities continue to control areas that are rightfully under their jurisdiction. The AHRI appears to be the program of a President who believes Washington, D.C., knows best and can govern best every aspect of life in every American community.

## Similar Objections

The Clinton Administration claims that the AHRI will help "reinvent government." But President Clinton's understanding of reinventing government seems to mean creating additional layers of bureaucracy. The American Heritage Rivers Initiative, in fact, is similar to an existing program, the National Rural Development Partnership (NRDP) established by President George Bush in 1991 by executive order. The NRDP is a flawed program: President Bush had no congressional authority over water rights, property rights, or the appropriation of funding when he initiated it; therefore, it also violates a number of constitutional provisions.

Like the AHRI, the NRDP planned to create a collaborative relationship among federal, state, local, and tribal governments, and private, nonprofit, and community-based organizations within each state and some territorial areas, in order to establish a comprehensive and strategic approach to rural development efforts in each state. A comparison of the descriptions of these programs from their respective World Wide Web sites reveals further similarities.

According to the Web site of the National Rural Development Partnership, the NRDP's objectives are to:

- *Encourage and support* innovative approaches to rural development and more effective resolution of rural development issues;
- *Develop* innovative approaches;

- *Build* partnerships among federal, state, local, and tribal governments and the private sector;
- *Encourage* local empowerment;
- *Involve* the Departments of Agriculture, Commerce, Defense, Energy, Housing and Urban Development, Interior, Justice, and Transportation, the Environmental Protection Agency, and the Army Corps of Engineers; and
- *Use* existing federal personnel and funds to work with the states to bring public and private resources together for solutions to local problems.

According to the Web site of the American Heritage Rivers Initiative, the AHRI is supposed to:

- *Encourage* community revitalization by providing federal programs and services more efficiently and effectively;
- *Develop* strategies that lead to action;
- *Build* a partnership between federal, state, tribal, and local officials, as well as private for-profit, nonprofit, and community-based organizations;
- *Encourage* community-led efforts;
- *Involve* the secretaries of the Departments of Agriculture, Commerce, Defense, Energy, Housing and Urban Development, Interior, and Transportation; the attorney general; the administrator of the Environmental Protection Agency; and the chairs of the National Endowment for the Arts, the National Endowment for the Humanities, and the Advisory Council on Historic Preservation; and
- *Use* existing federal staff, resources, and programs to assist communities.

## Creating Additional Bureaucracy

Reinventing government usually does not imply duplicating a federal program already operating in 38 states that has the same objective: promoting community involvement and development. Besides sharing the NRDP's objective, the AHRI will create three new costly layers of bureaucracy. The AHRI:

1. Creates an American Heritage Rivers Interagency Committee that will be responsible for implementing the AHRI.
2. Establishes a panel to review the river nomination packets and recommend rivers to the President for designation. The panel will include representatives from natural, cultural, and historic resources concerns; scenic, environmental, and recreation interests; tourism, transportation, and economic development interests; and industries such as agriculture, hydropower, manufacturing, mining, and forest management.

> *"It is difficult to comprehend how creating another federal program . . . will facilitate an efficient method of cleaning up America's great rivers."*

3. Gives the Interagency Committee the authority to transfer funds from other legitimate and congressionally authorized federal programs to fund ten

new river navigators appointed by the President. The new bureaucrats would be paid approximately $100,000 each year to assist officials in the ten communities selected by the President to locate existing federal programs and money that would be used to improve their waterfronts and rivers. Funds also would be transferred to compensate engineers, biologists, and foresters who would provide studies and expertise in implementing the initiative. The salaries of the river navigators would cost $1 million per year (which would be compounded annually because ten new river areas would be designated per year), and the cost of the engineers, biologists, and foresters would be added to the already estimated $4 million annual cost of the program. It is unclear whether such authority on the part of the Interagency Committee is a violation of the Spending Clause in Article I of the Constitution because the Spending Clause gives Congress—and only Congress—the power and authority to "draw [monies] from the Treasury."

President Clinton is planning to implement the AHRI at a time when the country is clamoring for Congress to downsize the federal government and give more control back to the states. The true definition of reinventing government is to make government smaller and more efficient. It is difficult to comprehend how creating another federal program—and one that is similar to an existing program—and adding new layers of federal bureaucracy will facilitate an efficient method of cleaning up America's great rivers. Secretary of the Interior Bruce Babbitt, in a recent speech entitled "United by Waters—How and Why the Clean Water Act Became the Urban Renewal Act That Actually Works," stated:

> Finally in 1972 Congress enacted a new law . . . [t]he Clean Water Act proclaimed a simple if awkwardly stated goal: make the nation's rivers, lakes, and shores "swimmable and fishable." As American cities used the Act to clean up and restore their waters, those waters, in turn, have begun to heal and restore our American cities.

Even as the Clinton Administration touts the effectiveness of the Clean Water Act in restoring and protecting American rivers, it boldly declares that the country also needs the AHRI. If Secretary Babbitt believes the goals of the Clean Water Act already are being achieved, then one must ask: What is the real reason behind the Clinton Administration's new initiative? . . .

At a time when the country wants to downsize government and revitalize the importance of the Tenth Amendment, and Congress is recognizing the necessity of empowering local communities and states even more, the American Heritage Rivers Initiative chooses the wrong approach for preserving some of America's great resources, its many rivers.

# The Efforts of Citizens Can Reduce Water Pollution

## by Donella Meadows

**About the author:** *Donella Meadows is an environmental writer and an adjunct professor of environmental studies at Dartmouth College in Hanover, New Hampshire.*

The good news is that our Clean Water Act, plus billions of dollars in municipal treatment plants and industrial wastewater processing, has rescued many of our streams and lakes from sewerhood.

The bad news is that, with the nastiest waste pipes cleaned up, we still insult our waterbodies with filled-in wetlands, runoff from lawns and farms, here a dam, there a dam, everywhere a little acid rain or toxic fallout. Ponds cloud up with strange weeds. Almost all the oysters are gone from Chesapeake Bay. Only one percent of the natural wetlands of Iowa remain. Warnings about contaminated local fish or shellfish are posted in 45 states.

### Laws Are Not Enforced

Water quality and water creatures continue to decline not because we lack protective laws, but because the laws are tepidly enforced. A report from the Environmental Defense Fund blames "inadequate authority, funding limitations, and bureaucratic timidity."

Within that bad news, however, there is a bit of good news. Where government fails, caring citizens are stepping in. In just a day of calling around New England, I uncovered a wealth of citizen efforts to monitor, protect, and restore local lakes and rivers. They're scattered, they're vastly underfunded, but they demonstrate how public and private efforts could join to clean up our water.

Watershed Watch at the University of Rhode Island, for example, keeps 250 volunteers busy at 90 locations taking weekly water samples from April through November. The university trains the volunteers, gives them the equipment, does the lab tests, and collates the information (which is passed back to the volunteers and on to the state Department of Environmental Management).

Reprinted, by permission, from Donella Meadows, "Watershed Watch," *Yes! A Journal of Positive Futures*, Fall 1996. Subscriptions $24/yr., single issues $8 (includes postage). P.O. Box 10818, Bainbridge Island, WA 98110. Telephone: (206) 842-0216. Fax: (206) 842-5208. Subscription information: (800) 937-4451. http://www.futurenet.org

Not only do students, auto mechanics, police officers, teachers and retired folks provide free labor, they also raise much of the amazingly low cost of the program, through their local conservation commissions or lake associations.

No one has to beat the bushes to find these volunteers. As word spreads, more communities ask to join. Says Elizabeth Herron of Watershed Watch, "It's great to have to go out on your favorite lake once a week. Hey, I'm not going fishing, I'm going monitoring!"

## Reliable and Professional Volunteers

Her job is to check the quality of the information coming in. She's delighted to report that well-trained citizens collect data as reliably as professionals. "You don't have to be a scientist. You just have to love a stream and be willing to follow directions."

Jody Conner at the Lakes Program of the New Hampshire Department of Environmental Services works with 500 volunteers to monitor 125 lakes. "I can't say enough for the volunteers," he says. "They're my ears and eyes. If there's a problem on the lake, they call me."

As in Rhode Island the New Hampshire program started with just one group asking for help in understanding just one lake. The next year there were 10 lakes, and things exploded from there. Now Conner's program helps people form lake associations,

> *"[Citizen efforts are] vastly underfunded, but they demonstrate how public and private efforts could join to clean up our water."*

puts out videos and books and kids' programs, and trains monitors to take water samples and, in advanced courses, to keep track of water plants, bugs and fish.

Each group gets an annual report with time graphs of how its lake is changing. The ones who've been in the program longest are beginning to trace water quality up tributaries. The New Hampshire volunteers also turn out information of professional quality. It goes into the state report on which waters are swimmable and fishable. It is helping a study of mercury accumulation in fish. Above all, it tells local governments and citizens when there is trouble in their water, so they can do something about it first hand and right away.

## The Work of Paul Godfrey

One of the most active and enthusiastic coordinators of citizen efforts on behalf of water is Dr. Paul Godfrey of the University of Massachusetts in Amherst. He started in 1983 helping citizens monitor acid rain throughout the state.

"One year we surveyed 200 lakes and streams for $100 each. The same year EPA did 2,000 waterbodies and spent $8,000 each," he said.

Now Godfrey holds together the Massachusetts Waterwatch Partnership, a coalition of state agencies, universities and citizens' groups on 50 lakes and 10 rivers.

Pressed for funds ("I spend 20–30 percent of my time looking for money"),

Godfrey has converted his garage into a workshop where he turns out low-cost monitoring equipment. He makes samplers for bottom water out of Mason jars, cement, epoxy, rope, and tubes from Bic pens. He produced 100 Secchi disks at half the cost of buying them. (Secchi disks look like dinner plates painted black and white in opposite quadrants. To measure clarity you lower them in the water to see how far down they are visible.)

## Citizens as Environmental Monitors

These programs—and there are many others—live hand-to-mouth. They don't have the budgets to cover all the waterbodies that need to be watched. They depend too much on a few dedicated experts who maintain quality and enthusiasm. And they focus on monitoring, which is not fixing.

But without monitoring, you don't know what needs fixing. And these citizen programs do much more than provide free information to governments too chintzy to fund the implementation of their own laws. They educate people about how water works, how important it is, and precisely how it gets messed up. They give folks the information and power to insist that government do what needs to be done. And they depolarize discussions, as more and more citizens know and understand the facts. Paul Godfrey saw this happen with his acid rain monitoring program, as the discussion changed from angry ideological stand-offs to "more like a conversation over the back fence."

You'd think state and federal agencies would know a good deal when they see it and fund these monitoring efforts properly.

Maybe sometime soon they will.

# Natural Remedies Can Be Used to Clean Up Toxic Waste

by Kara Villamil

**About the author:** *Kara Villamil is a science writer and media relations representative at Brookhaven National Laboratory in Upton, New York.*

As the twentieth century draws to a close, the United States can look back on its scientific and technological achievements with pride. Especially during the past 100 years, this country has been leading the world in both industrial and military strengths, and the end seems nowhere in sight.

## Technology Has Hurt the Environment

But these accomplishments have come with a price. One dimension that has suffered seriously is the environment. The American landscape is pockmarked with sites polluted by factories and power plants, mines and weapons labs. Many toxic chemicals and radioactive wastes linger on, bearing silent witness to the cavalier attitude with which we have treated our environment.

At government sites, the contamination remains in huge pits or inadequately lined underground tanks. At numerous shipyards and factory sites, it's littered throughout. At trash incinerators, the refuse of our consumer culture lies converted into piles of toxic ash. And America is not alone—the drive toward industrial and military development has left its marks around the world.

Recognizing the magnitude of the problem, the United States has set about to search for solutions. Since the early 1970s, much public scrutiny, a sea of regulations, and serious research have been brought to bear to make amends for our history of environmental neglect. Today, dozens of rules and regulations apply to different types of wastes, depending on the number of toxins they contain, the relative hazards of their constituents, and the amount and intensity of their radioactivity. Generally, the greater the hazard, the higher the cost of its disposal will be.

Kara Villamil, "Gentle Remedy for Harsh Pollutants," *The World & I*, September 1996. Reprinted with permission from *The World & I*, a publication of The Washington Times Corporation, Copyright ©1996.

Research has yielded some promising advances. One of them, developed by A.J. Francis and Cleveland Dodge at the Brookhaven National Laboratory in Upton, New York, is a surprisingly simple technique that employs bacteria, citric acid, and sunlight to remove toxic and radioactive metals from soil and sludge. Various tests have demonstrated that it can be highly effective. For instance, at one site in Ohio, it removed over 90 percent of uranium from heavily polluted soil. More tests have suggested that this three-part remedy may be applied to other nuclear and nonnuclear wastes, including incinerator ash.

> *"The American landscape is pockmarked with sites polluted by factories and power plants, mines and weapons labs."*

The U.S. Department of Energy (DOE) has inherited the responsibility for cleaning up much of this country's nuclear mess. Over 130 facilities, including sites as historic as Los Alamos and as obscure as one called Y-12, fall under the umbrella of the DOE. In many of these areas, the environmental legacy of the Cold War lingers in trenches, landfills, ponds, heaps, holes, and pits.

Luckily, the DOE also has at its disposal the scientific resources of its handful of university-sized research laboratories, one of which is Brookhaven. Several of these labs are researching methods by which the DOE can clean up its nearly four million cubic yards of buried waste contaminated with radioactivity.

Seven DOE facilities share the dubious distinction of having this country's largest landfills contaminated with nuclear waste. Because regulations before the 1970s were very lax, these sites became dumping grounds for both toxic chemical wastes and radioactive wastes. This practice, now banned, resulted in "mixed wastes," multiplying the challenge of cleaning them up. The volume of waste was further increased by "clean" trash that was later dumped in the same landfills.

## Dangerous Bacteria

Adding to these complications has been the activity of anaerobic bacteria in the soil where the waste is buried. The metabolism of these bacteria is well suited for life underground because they can use nitrate, sulfate, organic compounds, and metals found in the soil instead of oxygen.

Ever searching for food sources, these microbes are breaking down the hazardous wastes. They have a great capacity for reacting with radioactive metals like uranium and strontium, which are major components of buried nuclear waste. The "bugs" also attack heavy metals—such as chromium, cadmium, and lead—and hazardous organic chemicals, such as solvents and PCBs (polychlorobiphenyls).

Through these processes, known as *biodegradation* and *biotransformation*, the bacteria may alter the solubility of the metals in water, making it easier for the pollutants to be washed by rainwater into the surrounding soil and toward water supplies.

In light of the ability of soil bacteria to release hazardous waste ingredients, and the situation that underground containers are aging and leaking, cleanup is a race against time. Ironically enough, the innovative technique developed by Francis and Dodge uses one of the many species of soil bacteria that threaten the wastes buried in landfills, but it puts the microbes to work in a positive role.

## Citric Acid Treatment

Before embarking on this project, Francis, a microbiologist, had studied the basic processes of soil bacteria, particularly those that feast on citric acid. This simple acid is the same substance that gives oranges and lemons their tang. Each molecule of this acid—composed of carbon, hydrogen, and oxygen atoms—is small, but its structure makes it useful in a chemical process known as *chelation*, by which it captures a metal and holds it in place by forming a ring-shaped structure that includes the metal.

By the late 1980s, citric acid's chelating properties had begun to be used to wash radioactive contamination from nuclear reactor components. The acid was also seen as a potential cleaning agent for contaminated soil, such as that at DOE waste sites. The method consists of mixing a quantity of the soil with a solution of citric acid in water, allowing the acid to chelate the toxic and radioactive metals, and then separating the solution (containing the metals and citric acid) from the soil.

> *"In light of the ability of soil bacteria to release hazardous waste ingredients . . . cleanup is a race against time."*

This approach is more benign than earlier methods, which involved scrubbing contaminated soil with powerful chemicals like hot sulfuric acid or sodium hypochlorite. Because soil is a miniature ecosystem—including bacteria, fungi, minerals, and organic nutrients—treatment of the soil with such harsh chemicals not only removes the pollutants but also leaves the soil barren. In addition, such treatments generate witches' brews of their own—mixed waste solutions that have to be disposed of carefully and expensively.

By contrast, the citric acid treatment does not harm microbes in the soil. Nonetheless, it also produces a mixed toxic-radioactive waste solution that needs careful disposal at considerable expense. In light of this, Francis saw the need to combine the citric acid process with a means to separate the radioactive and toxic pollutants from the citric acid to reduce the amount of waste.

Recognizing the connection between the use of citric acid in waste treatment and his own research on the metabolism of citric acid and metals by soil bacteria, Francis teamed up with Dodge, a chemist. If the citric acid were used to chelate the radioactive and toxic metals at the waste sites, they reasoned, then the bacteria they studied could digest the bonds between acid and metal, allowing the metal to be recovered.

Who could foresee where their research would lead? Separating toxic metals from the radioactive ones would solve much of the problem of mixed-waste disposal—the soil and sludge could be separated from their pollutants with less expense and fewer byproducts. Perhaps the nonradioactive metals could be recycled to industries that could use them, and the radioactive elements could be packaged in impenetrable drums and stored in carefully managed waste sites.

To test these ideas, Francis and Dodge chose the bacterial species *Pseudomonas fluorescens*, a soil microbe originally isolated from a low-level radioactive waste site. Preliminary tests on simulated wastes gave promising results, but how would it work on actual waste?

The researchers obtained a sample of contaminated sludge from the uranium processing site called Y-12, in Tennessee. They treated it with citric acid solution, collected the yellow acid-metal brew in a flask, and mixed in a few million of the single-celled *P. fluorescens*.

The lowly bacteria launched into a flurry of activity. With a silent swiftness, they broke the chemical bonds between acid and metal and ate away the citric acid. The reaction, which generated heat, caused the metals to precipitate to the bottom of the flask, while the citric acid was converted to carbon dioxide and water. The precipitated metals could then be separated by centrifugation.

## Working with Uranium

All of the metallic pollutants, except uranium, were recovered in this manner. This is because most of the metals—including cobalt, chromium, manganese, nickel, strontium, thorium, and zinc—form a complex with citric acid that the bacteria can degrade. But uranium is different. Each molecule of citric acid can chelate two uranium atoms, not one. The bacteria are unable to break down this complex fully.

As it turns out, this difference works to the researchers' advantage: If radioactive uranium can be separated from the other metals, the mixed waste could be turned into two distinct wastes with lower disposal costs. But this led to the next question: Would it be possible to separate the uranium from its unique chelated form?

The answer was yes. Francis and Dodge found that a light-stimulated reaction, known as *photodegradation*, would do the job. When the citric acid-uranium solution was merely exposed to light in the presence of atmospheric oxygen, the bonds between citric acid and uranium were broken and the uranium precipitated to the bottom of the flask.

*"What was a pollutant in the ground can be a valuable component when separated out."*

In order to determine the detailed structure of the uranium precipitate, Francis and Dodge simply carried the sample across the street to another of Brookhaven's scientific facilities, the National Synchrotron Light Source. The Light Source accelerates electrons around a ring, generating intense beams of X rays and ultraviolet light. Researchers use these beams to analyze such things as crystals of proteins or materials in computer chips.

At the Light Source, Francis' team analyzed the uranium precipitate by a technique known as *extended X-ray absorption, fine structure*—a newly developed approach that is excellent for determining the structures of small molecules. In addition, they examined their sample at Professor Clive Clayton's X-ray photoelectron spectroscopy laboratory at the nearby State University of New York at Stony Brook. The combined results revealed that the precipitate consisted of uranium oxide. Each molecule was composed of one uranium atom bound to three oxygen atoms, and was complexed with two water molecules. This finding helped guide the researchers in selecting a method to dispose of the uranium.

> *"This combination of citric acid, bacteria, and sunlight appears to be a promising weapon in the arsenal against environmental pollution."*

## The Advantages of This Method

This three-part remedy of citric acid, microbes, and sunlight offers a number of advantages. First, the citric acid process avoids the disastrous effects on the soil that marred the earlier use of highly caustic or acidic chemicals. Instead, it allows the metals to be removed without upsetting the soil's fragile ecosystem.

Second, citric acid-metal chelates can be readily degraded by the *P. fluorescens* bacteria. These microbes convert the citric acid to plain old water, carbon dioxide, and heat—no nasty byproducts to contend with.

Moreover, the metals recovered by the process can potentially be recycled by industry. What was a pollutant in the ground can be a valuable component when separated out. In the case of uranium, the process separates it from the other metals and allows it to be precipitated in the oxide form by simple photodegradation. But its radioactive properties make its careful disposal a necessity. The DOE operates several waste storage facilities where radioactive waste is secured in many layers of packaging and kept under monitored conditions until a permanent disposal facility can be built.

The Brookhaven researchers have already obtained several U.S. patents. But their method requires further testing before it can be fully transferred from the laboratory to the contaminated waste sites.

In an initial test on sludge from the Y-12 plant in Tennessee, the citric acid removed 87 percent of the uranium in one waste sample. Francis and Dodge then tested their method on soil from another site in Ohio, which has over two million cubic yards of uranium-contaminated soil. The citric acid was shown to remove 90 percent of the uranium from a sample of the Ohio soil.

The process also removed other metals effectively. The biodegradation led to the recovery of nearly all of the cobalt, chromium, copper, strontium, thorium, and zinc; more than 90 percent of the lead; and more than half of the nickel.

The scientists also saw the potential for the process to be useful in treating other wastes, such as the ash produced in trash-burning incinerators. Because it is the end product of burning comingled trash—including batteries, paints, and bits of metal—this ash often has high levels of heavy toxic metals. This makes it hard (not to mention controversial) to dispose of the ash in a landfill—there's always the possibility that the toxics will leak from the landfill and into neighboring groundwater. If the toxic metals can be removed, the ash will be much more benign.

Francis and Dodge have found an industrial partner that specializes in the processing of incinerator ash. The company has agreed to work with them to test the feasibility of their method with this type of waste. That trial, though not complete, already shows promising results.

Nonetheless, the transition from lab to "real world" is not a swift one. Francis and Dodge hope to improve their method by optimizing the biodegradation and photodegradation steps. Before the technique can be used on a large scale, a method for recovering and recycling any leftover citric acid needs to be developed. In addition, because the composition of waste can vary greatly even at different locations of the same site, the two scientists are continuing their tests with a variety of samples.

All in all, this combination of citric acid, bacteria, and sunlight appears to be a promising weapon in the arsenal against environmental pollution.

Its story is one full of ironies: One DOE lab works to help clean up others; one type of soil bacteria safely removes pollutants that other soil bacteria would unleash. But perhaps the greatest irony of all is that a simple, all-natural technique can tackle artificially generated hazardous wastes that were once discarded with little regard for their environmental impact. Let us hope that our ignorant abuse of nature can still be remedied.

# Chapter 4

# Can Free-Market Approaches Protect the Environment?

CURRENT CONTROVERSIES

# Chapter Preface

Environmental activists and regulators are often at odds with business leaders, who believe that efforts to conserve the environment sometimes destroys jobs and profits. Rather than the use of government regulations to protect the environment, corporate leaders advocate a free-market approach. They believe that if government gets out of the way, industries will conserve the environment because it is in their economic interest to do so. The controversy over how best to preserve forests exemplifies the debate between supporters of government regulations and proponents of free-market solutions.

Many environmentalists claim that forests need to be protected through the actions of the U.S. Forest Service and government measures such as the Northwest Forest Plan, adopted in 1993, which limited logging on most federal lands. They view logging as a threat to the biodiversity of these forests. According to Paul and Anne Ehrlich, logging depletes the supply of fish because the runoff from logged areas damages the streams and rivers where the fish live. Logging also leads to "lost esthetic values and diminished ecosystem functioning," according to the Ehrlichs.

Others assert that logging protects the environment. Dennis T. Avery, a senior fellow at the Hudson Institute, a public policy research organization, argues that logging mature trees is beneficial because it reduces the risks of forest fire. When done properly, logging may increase species diversity. For example, making small clear-cuts in old-growth forests lets in light and opens space, attracting more wildlife. He argues that companies that log only diseased or less-valuable trees end up losing profits and cutting jobs.

A balance between a thriving economy and a healthy environment may be difficult to achieve. In the following chapter, the authors consider whether free-market approaches are the best way to conserve the environment or if a reliance on business is unwise.

# Free-Market Environmentalism Can Protect the Environment

**by Collette Ridgeway**

**About the author:** *Collette Ridgeway is the marketing assistant at the Institute for Humane Studies at George Mason University in Fairfax, Virginia.*

Everybody has heard of Yellowstone National Park and even Tongass National Forest, the former under the jurisdiction of the National Park Service and the latter under that of the Forest Service. But how many people have heard of the North Maine Woods, the Maine Coast Heritage Trust, or the Eastern Pennsylvania Hawk Mountain Sanctuary, all of which are under nongovernmental control? Those and many other private parks, wildlife sanctuaries, and nature reserves are doing a much better job of preserving natural amenities than are the federal agencies.

Because of our shared interest in both environmentalism and liberty, my husband and I spent a year-long "working honeymoon" crisscrossing the country visiting private nature preserves and documenting their exciting and vitally important story. I focused on the writing, while my husband Scott, a professional fine-art photographer, recorded on film the lands that were quietly being preserved by individuals, conservation organizations, and business enterprises.

## Free-Market Environmentalism

A year earlier, while studying geology at Pennsylvania State University, I had taken a course entitled "State of the World," after the annual publication of the same name. After many months of listening to unconvincing eco-statist rhetoric in class, I decided to do my research project on how free markets and private property rights could lead to more efficient and more permanent solutions to the many environmental problems facing the world today.

When I first presented my ideas in class, I found that most of the other stu-

Reprinted, by permission, from Collette Ridgeway, "Privately Protected Places," *Cato Policy Report*, March/April 1996.

dents had never heard of a free-market approach to environmental policy. A few were skeptical; some were intrigued; but most were simply surprised. As my research progressed, it became clear why most of my classmates had never been exposed to free-market environmentalism. Even though I was actively looking for information on alternatives to government control, I found it difficult to locate the resources I needed. When I did find them, they were hardly as exciting as the message put out by the traditional environmental groups. Much of the literature on free-market environmentalism is couched in technical economics terms, such as externalities, transactions costs, opportunity costs, and capitalization. It was difficult for my classmates—as it seems to be for most other people—to visualize how a market system based on property rights, contracts, and the rule of law can safeguard environmental amenities better than a command-and-control governmental system.

To bring free-market environmentalism to the attention of more people in an emotionally appealing way, Scott and I decided to produce a coffee-table photography book showing the lands of private owners as protected places rather than as property requiring government protection. We contacted the Political Economy Research Center (PERC) and the Competitive Enterprise Institute (CEI), two leading free-market environmental groups, and got suggestions and a modest (but very much appreciated) grant to allow us to visit and photograph numerous privately protected areas around the country.

Although the experts at PERC and CEI offered some suggestions about

*"Private parks, wildlife sanctuaries, and nature reserves are doing a much better job of preserving natural amenities than are the federal agencies."*

places to go, there is no central listing of the myriad privately preserved places. Often owners have no desire to open their property to visitors, because their concern is to minimize the damage too many visitors cause (witness Yellowstone or other state-owned parks). Some privately preserved places, such as those owned by private conservation organizations or businesses, are better known than others, so we used them as a starting point.

Our first home as husband and wife was a 1987 Dodge van we had converted into a makeshift camper. It had neither air conditioning nor running water; but it did have a propane stove and a good stereo, and it was already well broken in with an odometer reading of 115,000 miles. We added to those miles as we wound our way from Maine to New Mexico and from Oregon to Pennsylvania.

## Protecting Maine's Environment

We decided to start in the northeast and follow the autumn color back to Pennsylvania, so we made our way to Maine. Our first destination was the North Maine Woods, almost 3 million acres of working forest. Although not a true wilderness, the woods have remained essentially the same as they were a century ago, with vast tracts of forest and lakes, some accessible only by canoe or floatplane.

A cooperative effort joining property owned by 25 different timber and paper companies, the North Maine Woods are accessible through just a few user-fee stations. There are no visitor centers with park rangers, no gas stations or even paved roads. It is one of the few places where you can find true solitude. We would often spend days at a time with only moose, black bears, and beavers for company.

> *"A market system based on property rights. . . can safeguard environmental amenities better than a command-and-control governmental system."*

Next we made our way to the Atlantic coast to see some of the properties owned by the Maine Coast Heritage Trust. Lobsterman Jasper Cates lives along one of the most beautiful and pristine sections of Maine's rugged historic coastline. When developers threatened to build homes on one particular portion of rocky shoreline, Cates realized that the best way to protect it was, not to seek government intervention, but to buy the land. Gathering funding and support, he and a handful of concerned neighbors founded the Maine Coast Heritage Trust to purchase and protect what they could of Maine's remaining undeveloped rocky coast. What began as a plan to protect the historical and environmental treasures of a small community now protects thousands of acres of pristine islands and coastal property all along Maine's Atlantic shore.

We also visited Maine's Acadia National Park. Although the park is quite beautiful, it was hard to ignore the roads, cars, countless people, and accompanying development. The property of the trust, on the other hand, looked just like it had 200 years and more ago, thanks to Jasper Cates and his neighbors.

As we left Maine and made our way south to New York State, we began to learn more and more about the Nature Conservancy, one of the most respected environmental organizations. As we talked with people, we learned that the conservancy is not always what it seems.

## A Threatening Conservancy

Many private landowners in the Adirondack Park region of New York State, such as David Howard, founder of the Land Rights Foundation, view the Nature Conservancy as a threatening entity set on purchasing the private land inside the "green line" borders of the park. Because of onerous edicts and bureaucratic commands, many small private landowners within the park boundaries feel pressured to sell their property to the state for inclusion in the park. Because of continual and punitive government harassment, many decide to sell to the Nature Conservancy instead. What they don't realize is that often the conservancy will turn around and sell their family property to the government, often at a very great profit.

With a new wariness about the intentions of some conservation organizations, we made our way to Pennsylvania and another privately owned preserve. The Hawk Mountain Sanctuary in eastern Pennsylvania lies along a natural migration route. In the early part of this century though, it was the end of the migration for thousands of raptors each year. At the time, all predators were considered undesirable. Hunters would stand on the rocky cliffs and shoot hundreds of birds a day as they slowly spiraled on the thermal updrafts.

Rosalie Edge, a leading conservationist of her day, finally stopped the slaughter by purchasing the mountain in 1934. Because of Edge and her sanctuary manager, Maurice Brown, both of whom literally risked their lives to prevent hunting of the birds they treasured, bird watchers now gather where hunters once stood. Today the 2,000-acre reserve is considered one of the best birding locations in the world.

## Ranchers and Conservation

After a brief rest at home over Christmas, Scott and I headed west on the next leg of our journey. This time we planned to visit lands owned by individuals and families, rather than by organizations or businesses as we had in the East.

When we approached ranchers to explain our project, we were greeted with gratitude and hospitality. We were welcomed into their homes, communities, and families. Most of them never get credit for the conservation and restoration work they do on their land. They were happy to, for once, be seen as friends of the environment.

We visited many landowners on the trip, but a few stand out. The Yamsi Ranch and Iram Wild Horse Preserve are both run by the Hyde family. Dayton Hyde is perhaps best known for his many books about his experiences on his Oregon ranch and for breaking a seemingly unbreakable rancher's law—he welcomed coyotes on his land. Instead of killing them, he fed them. As a result, Hyde no longer needed to worry about coyotes attacking his herds.

Today, Dayton's wife, Gerda, and their grown children manage the Yamsi Ranch and continue the family tradition of holistic planning. On the afternoon we arrived, son John and his brother-in-law Scott Jayne were following in Gerda Hyde's footsteps by replanting willows along the banks of the Upper Williamson River. Over the years, Gerda herself replanted, by hand, over 250,000 trees on the ranch.

The Hydes have also made other improvements to their property. John Hyde explained to us the necessity of cattle in a brittle, arid environment,

*"Most [ranchers] never get credit for the conservation and restoration work they do on their land."*

such as theirs. He showed us how, without grazing, the live grasses become choked with clumps of dead plant matter until the entire plant eventually dies. To keep the grasslands green, the Hydes graze each parcel intensively for a short period of time, followed by a long rest to allow the grasses to regrow.

Near the end of our tour of the property, John Hyde drove us to a natural spring that empties into the river. As we drank from the spring, John told us that, unlike many ranchers, they use no pesticides on their cattle that might contaminate this pure water. The Hydes have controlled the fly problem by selective breeding.

> *"Intentional destruction and vandalism of natural treasures are . . . unavoidable problems on public property."*

Although the Hydes have done a great deal of work on their own property, their influence can also be felt elsewhere. In addition to starting the Wildlife Stronghold program, an association of landowners dedicated to protecting the wildlife on their property, the Hydes took action to protect the delicate watershed surrounding their ranch. A five-year plan was worked out that brought together an interesting mix of Native Americans, ranchers, liberals, and conservatives to protect the watershed of the Upper Williamson River. Just as the program looked as if it might be a success, Forest Service regulations brought all efforts to a halt. Nothing could be done to the watershed that was on National Forest land. Without that vital part, all work was in vain.

Undeterred, the Hyde family continues to reach out to educate the community. By conducting school field trips, trail rides, and offering catch-and-release fly fishing opportunities, the Hydes make sure that word of their work spreads.

Lately, Dayton Hyde has turned his attentions to another project, the Iram Wild Horse Sanctuary, in the Black Hills of South Dakota. The sanctuary, founded as a nonprofit organization, provides a home for unadoptable wild horses on over 11,000 acres of spectacular land that are also home to scattered remains of Native American tepee rings and petroglyphs.

Dayton Hyde knows that public property rarely serves the best interest of the environment. Many National Parks face the constant problem of preventing the great numbers of visitors from damaging what they have traveled so far to enjoy. Trail erosion, litter, and even intentional destruction and vandalism of natural treasures are all unavoidable problems on public property. That "tragedy of the commons" does not occur on protected private property. Hyde told us that no additional damage has been done to the petroglyphs, or to any other part of the sanctuary, since it has been in his care. Because access to the horses' habitat can be carefully controlled, the land and the horses can thrive.

Guided tours, trail rides, special western events, and overnight accommodations in authentically styled tepees attract many visitors to the sanctuary. Tourism is the driving force that allows the sanctuary to remain in operation.

As our journey continued farther south, we met a man who has brought new life to a ruined and dying landscape in the mountains of central New Mexico. Thirteenth-century petroglyphs of fish and beaver in the area gave a very different picture of the dry and barren tract of land that rancher Sid Goodloe had purchased in 1956. Through 40 years of hard work, Goodloe was able to successfully rebuild the natural savanna ecosystem historically found there.

By a combination of cutting, bulldozing, and prescribed burning, Goodloe removed water-hungry piñon and juniper trees that had taken root as a result of decades of overgrazing and unnatural, government-sponsored fire-suppression programs. His goal was to return the land to its natural state. Like the Hydes, he used his cattle as an environmental tool.

Whereas the natural fire cycle formerly removed excess brush, Goodloe now does so by hand. Invading water-hungry piñon and juniper created an incredible demand on the area's typically minute water resources. Once the piñon and juniper had been removed, long-dead springs came alive and water began to flow in the streams once again. Careful fencing of the stream banks allowed natural willows and cottonwoods to take root as silt re-covered the eroded bedrock of the stream. Today, trout thrive in the small stream beside Goodloe's modest adobe house, just as they did many centuries ago.

Native grassland plants also began to thrive on the open prairie when the piñon and juniper were gone. Elk, mule deer, and turkeys now flock to the property from the surrounding National Forest lands to feed on the new bounty.

Ever since the original Smokey the Bear was found as a badly burned and injured cub just a few miles from Capitan, New Mexico, total fire suppression had been the unquestioned rule of the forest. Goodloe's dramatic success with his property, however, finally gave Forest Service managers reason to question that policy. Ironically enough, when local managers tried to make the same improvements to abused and overgrown National Forest property, all attempts at change were halted by the Endangered Species Act: the National Forest contained habitat of the endangered goshawk.

On July 3, 1994, as Capitan celebrated the 50th anniversary of Smokey's rescue, lightning struck the Patos Mountain National Forest bordering Sid Goodloe's property. Years of governmental fire suppression had allowed an unnatural buildup of dead brush that encouraged the flames to climb high into the upper canopy of the trees. The goshawk habitat, along with much of the surrounding National Forest property, was destroyed.

In the end, it was Goodloe's management practices that allowed the fire to be contained. The firefighters used as the edge of their backfire his ranch, which sustained only a mild cleansing burn that recycled the biomass nutrients; rid the area of underbrush; and left the larger trees, with their natural resistance to low-intensity fires.

When we started our trip, we had an intellectual grasp of free-market environmentalism. We knew about the incentives involved in private ownership and the economic realities of the market. We learned much more as we talked with, and worked alongside, individual landowners, the real environmentalists, who are motivated by a love and respect for the land—their land.

# Green Marketing Can Help Conserve the Environment

by Hector R. Lozada and Alma T. Mintu-Wimsatt

**About the authors:** *Hector R. Lozada is an assistant professor of marketing at Seton Hall University in South Orange, New Jersey. Alma T. Mintu-Wimsatt is an associate professor of marketing at Texas A&M University in Commerce, Texas.*

In the 1970s, leading marketing thinkers like Philip Kotler and Gerald Zaltman mandated that "social marketing" become an important concept in the discipline. Social marketing was defined as "the application of marketing concepts and techniques to the marketing of various socially beneficial ideas and causes instead of products and services in the commercial sense." This definition implicitly includes ideas on the preservation, conservation, and protection of the physical environment as a component of social marketing.

Building on the tenets of social marketing, Karl E. Henion and Thomas C. Kinnear offer a definition of *ecological marketing:*

> . . . [E]cological marketing is concerned with all marketing activities: (1) that served to help cause environmental problems, and (2) that may serve to provide a remedy for environmental problems. Thus, ecological marketing is the study of the positive and negative aspects of marketing activities on pollution, energy depletion and nonenergy resource depletion.

## Defining "Green Marketing"

More recently, Alma T. Mintu and Hector R. Lozada have defined *green marketing* as "the application of marketing tools to facilitate exchanges that satisfy organizational and individual goals in such a way that the preservation, protection, and conservation of the physical environment is upheld." Through this definition, Mintu and Lozada note that green marketing goes beyond image-building activities. The ecological concerns espoused by Henion and Kinnear would be integrated into the strategies, policies, and processes critical to the organiza-

Hector R. Lozada and Alma T. Mintu-Wimsatt, "Green-Based Innovation: Sustainable Development in Product Management," in *Environmental Marketing: Strategies, Practice, Theory, and Research*, edited by Michael Jay Polonsky and Alma T. Mintu-Wimsatt (Binghamton: Haworth Press). Copyright ©1995 by The Haworth Press. Reprinted by permission.

tion. More importantly, this definition of green marketing parallels what practitioners such as Walter Coddington are embracing as *environmental marketing:*

> Marketing activities that recognize environmental stewardship as a business development responsibility and business growth opportunity is what I mean by *environmental marketing.*

> The environmental marketer adds the environment to the standard mix of decision-making variables.

Thus, *green marketing,* as we use the term here, conveys a more proactive role for marketers. It fosters not only sensitivity to the impact that marketing activities may have on the natural environment, but also encourages practices that reduce or minimize any detrimental impact. . . .

## Corporate Solutions

The basic ideas behind environmentalism dictate that corporations have responsibilities that go beyond the production of goods and services. These responsibilities involve helping to solve important social problems, especially those they have helped create. Corporations such as McDonald's, Wal-Mart, Procter & Gamble, and Du Pont acknowledge that the environment must be protected and enhanced for economic growth to take place, and have taken action towards that goal. McDonald's has made a $100 million commitment to its consumers for recycling purposes. Wal-Mart encourages the purchase of environmentally friendly products and reports that the green labeling program that they initiated in 1989 contributed to an overall 25% increase in sales for the year. Procter & Gamble has pledged to spend $20 million per year to develop a composting infrastructure.

Yet, note that the Procter & Gamble example is quite telling. To a large extent, the company has been under fire by environmentalists mostly for its disposable diapers and its detergents. As a response, Procter & Gamble has implemented a strategy that takes the concepts of recycling and reusing to heart, particularly regarding packaging. Still, they have discovered that the synergistic relationship between issues and trends can yield criticism and consumer resistance. Even though their formula for Cheer laundry detergent (or Ariel outside of the U.S.) has been changed to minimize the amount of phosphates in the product, the company is still being strongly criticized for its overt reliance on animal testing.

> *"[Green marketing] fosters . . . sensitivity to the impact that marketing activities may have on the natural environment."*

In spite of some setbacks, green marketing efforts on the part of corporate America continue to grow. As a result of the re-greening of society, *environmentalism* has slowly become a buzz word for corporate policies of the 1990s. Within the context of corporate activities, environmentalism is interpreted as a higher level of corporate consciousness geared toward the protection, preservation, and conservation of the physical environment.

All in all, environmentalism could act as a trigger to innovation, one that is desperately needed and most welcomed. The activities associated with sustainable development may be construed by business either as a potential threat (increased regulatory action on the part of government, consumer resistance to or avoidance of environmentally unfriendly products), or as an opportunity to achieve their goals by doing what is right for the planet

> *"Integrating environmental and growth concerns presents new challenges for the long-term, for which creativity is imperative."*

and for society at large. However, Coddington forewarns that green-product development may not be an easy sell to senior management due to an apparent widespread perception that the introduction of green-product lines may have a negative impact on sales.

## New Opportunities

We firmly believe, though, that at a time when corporate America is coming to terms with the inefficacy of some of their short-sighted alternatives implemented in the pursuit of growth, green-based innovation may represent an invigorating alternative. The ability to launch new products that are internally developed provides more control over the desired growth that the corporation seeks, especially when a strategic window opens up. Today, integrating environmental and growth concerns presents new challenges for the long-term, for which creativity is imperative. Green products represent a substantial product opportunity, the opening of a provocative strategic window.

Take, for instance, the following corporate examples. The Royal Dutch/Shell Group is one of a growing minority of companies that are forming task forces, mounting experiments, and revising their planning based on the idea of balancing growth and the environment. Braden R. Allenby, AT&T's senior environmental attorney, suggests that corporate leaders are tiptoeing to a new approach to management that drives a different set of design and cost considerations. 3M had as a goal for 1995 to reduce air and water emissions by 90% and solid waste 50% from the levels of 1990. This will cut the inflation-adjusted cost per unit of most products by 10%. Monsanto, Du Pont, and AT&T, like 3M, are planning to sharply cut air emissions and waste. Dow Chemical may replace chlorinated solvents used for cleaning industrial equipment with less-polluting, water-based systems. BMW is using design for disassembly in a pilot project aimed at building cars so they can eventually be taken apart and recycled more easily. This concept is also being implemented by IBM-Germany to dismantle and recycle computer parts. All of these corporate attempts are not only meant to capitalize on the potential profitability and cost reduction associated with green-product development. [According to Coddington], it is also an opportunity "to improve their bona fides with a populace that is demanding exemplary environmental behavior from the corporate community."

Equally vital is the integration of environmental issues into all aspects of the corporation's activities, from strategy formulation, planning, and construction through production and into dealings with consumers. There is no doubt that environmental considerations will open new business opportunities in the development of new technology. On the product side, environmental problems also require creative and, perhaps, innovative solutions. For example, manufacturers of products sold as canned sprays (e.g., Gillette's Right Guard Deodorant Spray, or more recently, the introduction of the Cool Wave line) have been replacing chlorofluorocarbons (CFCs) in their products before the deadline for their elimination. Additionally, we are currently seeing the revival of some old product forms, usually with minor (cosmetic) modifications (e.g., the pump spray bottle).

Similarly, several products have been reintroduced to the market (e.g., Procter & Gamble's Downy) in more environmentally friendly forms. Chris Hampson contends that almost all business will in fact be affected by environmental considerations. Hence, firms' failure to be sensitive to these influences can make them lag behind in the competitive race.

## The Future of Green Marketing

Although some would prefer to look at these changing circumstances as threats, visionaries within business firms are realizing that there also are real opportunities in environmental developments for those companies ready to recognize and capitalize on them. To support our contention, we borrow from Frances Cairncross, who suggests that developing products that use nature more frugally will call forth whole generations of technology.

The change will be more pervasive than those that followed the invention of the steam engine or the computer. Fortunes await those who devise less expensive ways to dispose of plastics or to clean up contaminated soil. The great engineering projects of the next century will not be the civil engineering of dams or bridges, but the bio-engineering of sewage works and waste tips. Industry has before it that most precious of prospects: a spur to innovate.

# Ecotourism Will Protect the Environment

**by John Whiteman**

**About the author:** *John Whiteman is a partner in Whiteman and Taintor, a tourism and community-development consulting firm in Massachusetts and Colorado.*

Ecotourism is often touted as an experience that allows us a rare opportunity to have our cake and eat it too. More precisely, ecotourism allows us to enjoy wilderness experiences without allowing those experiences to compromise the overall health of the natural environment.

When properly managed, ecotourism offers excellent wilderness experiences while contributing to the preservation of natural and historic places. It works this way: Controlled numbers of people pay for the opportunity to visit a sensitive environmental area. While there, they enjoy a sense of spiritual renewal. And they leave behind an intact ecosystem and increased wealth for the local community.

## Defining Ecotourism

Ecotourism is a relatively new concept. As a result, its meaning continues to evolve.

In *Nature Tourism,* Richard Ryel and Tom Grasse, president and director of marketing, respectively, for International Expeditions, define ecotourism as:

> purposeful travel that creates an understanding of cultural and natural history, while safeguarding the integrity of the ecosystem and producing economic benefits that encourage conservation.

The Ecotourism Society, a non-profit organization based in Bennington, Vermont, broadens that definition by adding that the economics of ecotourism should be "financially beneficial to local citizens."

The Canadian Environmental Advisory Council's *Ecotourism in Canada* defines an ecotourism trip as one that:

Reprinted, by permission, from John Whiteman, "Ecotourism Promotes, Protects Environment," *Forum for Applied Research and Public Policy*, Winter 1996.

• Provides first-hand experience of the natural and/or cultural environment.

• Involves experiencing nature on nature's, not the visitor's, terms.

• Accepts that access to and use of natural and cultural resources must be limited.

• Includes grassroots involvement from the planning through the delivery of ecotours.

• Promotes environmental ethics and provides benefits to participants through education and interpretation.

• Offers economic benefits to the tourism industry.

• Directs a portion of the revenues to the maintenance and enhancement of the natural- or cultural-resource base.

## Ecotourism Has Evolved

With ecotourism organizations emphasizing such goals, the field is evolving in ways that make it vastly different from traditional tourism.

In the past, the tourism industry paid little attention to travelers' impact on natural resources.

Fortunately, that's changing largely because of pressure from communities, government, environmental groups, and tourists themselves. All are encouraging tourism operators to work within ecologically and culturally sustainable principles. They are also pressing operators to cooperate with a diverse range of partners, including state and federal agencies charged with the preservation of the natural and cultural environment.

## Ecotourism Research

Ecotourism research has focused primarily on the emerging ecotourism market in less-developed countries, where ecotourism began as a low-capital and low-tech economic-development strategy.

This situation is changing, however, as university researchers track ecotourism's growth in North America and explore the field from a variety of perspectives.

For instance, Paul Eagles of the University of Waterloo has examined what motivates Canadian ecotourists to vacation where they do. He found ecotourists to be high-energy "time maximizers," not laid-back "treehuggers."

University of Montana researchers Michael Yuan and Neil Christensen calculated that the economic impact of ecotourism is equal to or greater than the impact of traditional tourism. Despite such evidence, ecotourism has not received a great deal of funding from state and provincial governments—that is, until recently. Two reports, published in the early 1990s, reveal the increasing lure of ecotourism.

The first is the U.S. Travel Data Center's report, *Discover America: Tourism and the Environment,* which explores the role of environmental issues in shaping the decisions of today's travelers. The second consists of ecotourism market studies, conducted by the provinces of British Columbia and Alberta, to assess the status of ecotourism in North America. Specifically, these studies address the potential for these provinces to attract ecotourists from other parts of North America and Europe.

## Debating the Values of Ecotourism

While these studies are a start, the field of ecotourism in North America generally begs for additional research. U.S. and Canadian government tourism officials often bemoan the dearth of data on ecotourism. Many are concerned that a lack of reliable information has contributed to a lack of familiarity with the prospects for ecotourism among both tourists and tour operators.

Equally important, the absence of a consensus on what ecotourism means has left the term open to misrepresentation about its long-term value and impacts.

Some economically marginalized rural communities view ecotourism as a way to move their local economies from the red into the black. Other communities regard ecotourism as a veiled effort by environmentalists to control private land use and stymie economic progress.

For instance, Maine's Moosehead Lake region has struggled to balance the interests of the rafting industry with those of paper and timber companies. Paper companies have sometimes resisted ecotourism largely because they fear more environmental restrictions will follow on the heels of the initial safeguards.

> *"When properly managed, ecotourism offers excellent wilderness experiences while contributing to the preservation of natural and historic places."*

Tour operators are also confused about the term "ecotourism." Some insist they are drawn to ecotourism because they believe it attracts high-paying travelers. Others reject ecotourism, however, because they feel it attracts tight-walleted, low-budget, back-to-nature types who will add little to the local economy.

Such disparate views suggest that additional research is needed to assess the true effects of ecotourism on local economies and the travelers it attracts.

## Ecotourism and Education

When it comes to educational programs designed to enhance ecotourism's value, Canadian national and provincial governments are well ahead of those in the United States.

For instance, Canada's federal tourism agency, Tourism Canada, has prepared educational materials aimed at helping tour operators and local governments market ecotourism. And Alberta has published materials offering tour operators' and local governments' advice on promoting and managing wildlife-observation programs.

Although ecotourism educational activities in the United States lag behind those in Canada, there has been some progress. The Ecotourism Society has produced a publication offering ecotourism guidelines for U.S. tour operators.

## State Governments Are Important

State governments have historically played a leading role in promoting traditional tourism; some, however, have been more cautious in their willingness to support ecotourism.

In some states, tourism directors have left ecotourism promotion to local governments or the private sector because the market is small and remains highly segmented. Furthermore, tourism directors in these states often are reluctant to promote ecotourism because extensive advertising campaigns might attract more people than the natural resources can handle.

These states are the exception not the rule. Most state tourism departments are involved in examining or actively promoting ecotourism and have integrated ecotourism messages into their overall marketing program.

## Promoting Ecotourism

Oklahoma, for instance, blends promotion of its tall-grass prairie ecosystems into the state's overall "Native America" tourism campaign that integrates American Indian, European, and natural history. Meanwhile, small communities in the state's panhandle share management responsibilities for their cultural and natural areas with non-profit organizations that have recently acquired rare natural habitats. State government assists in promoting the area and linking these newly protected resources with nearby state-owned lands to broaden the overall experience.

The Oregon State Department of Tourism publishes a statewide wildlife-viewing guide. Nebraska promotes opportunities for tourists to witness the annual migration of the sandhill cranes and advertises canoe trips on the Niobrara River, a national scenic river. Washington state publishes seasonal guides directing visitors to outdoor experiences, including salmon fishing, wind-surfing, and whale watching.

While nearly every state and province has played a role in promoting environmental tourism, many states fail to strictly follow ecotourism management ethics.

> *"The economic impact of ecotourism is equal to or greater than the impact of traditional tourism."*

In part, the problem arises because some state tourism agencies are promoting ecotourism much as they have promoted other forms of tourism. These agencies also are understaffed, which makes managers reluctant to invest the resources necessary to adopt more time-consuming ecotourism management principles. In addition, some state tourism agencies must also tread a narrow political path to avoid the perception of being too closely aligned with the environmental movement.

## Creating Alliances

The U.S. National Park Service is one of the nation's oldest environmental re-source/tourism stewards, but the "one-owner, one-manager, one-marketer" model on which it is based is becoming increasingly rare on both the state and national level. In fact, many state resource-management agencies are creating alliances with non-governmental groups.

In Vermont, for instance, state and local governments have participated with private and non-profit groups to create the Moosalamoo Partnership, an ecosys-tem-based marketing and management project aimed at promoting 20,000 acres in southern Vermont. The centerpiece of this effort is the integration of many small businesses and agencies into a loosely federated organization that pro-motes the region without a central coordinating body.

More than 20 partners are active in the project. All share responsibility for marketing and managing the resources and tourism experiences within the eco-system.

This coordinated approach to promotion and management likely will become the norm because of (1) limited federal and state funding for ecotourism, (2) many resources cut across jurisdictional boundaries, and (3) ecotourism's grass-roots philosophy.

## Developing a Market for Ecotourism

With ecotourism's emphasis on local community development, more states are helping their cities and towns develop their ecotourism markets.

Nevada offers rural areas matching grants, for instance, to promote eco-tourism and develop the infrastructure necessary to handle the expected in-crease in ecotourists.

Tennessee offers tourism planning grants to help communities develop com-prehensive plans aimed at eco- as well as traditional tourism.

On the federal level, the Department of Housing and Urban Development makes rural economic-development funds available to communities to create strategies, including those focused on ecotourism.

Meanwhile, Federal Highway Administration funds are available to commu-nities hoping to develop corridor-management plans for promoting their road-side natural and cultural resources.

States have historically helped boost tourism and promote economic develop-ment by investing in such tourism facilities as conference centers and histori-cally oriented parks. A notable ecotourism example is the proposed Greylock Glen Ecotourism Resort, which will be situated on a 2,000-acre parcel of land surrounding Mount Greylock, the tallest peak in Massachusetts' Berkshire Mountains.

State and local governments are now guiding the tract, which over the past 20 years has shifted between private and public ownership. When completed, this site will boast conference, education, and recreation facilities, all with an "eco" theme.

In fact, local governments and the state's department of environmental management are now guiding designers as they begin plans for the resort, which will be privately financed and developed. The state has pledged $7 million for infrastructure development.

The resort will be built using sustainable, environmentally sensitive technology and will offer conferences as well as educational and recreational programs that pertain to environmental protection and management.

As an off-shoot, the state is developing an industrial park nearby that will focus on the research, development, and manufacturing of "eco-products" for use in the construction and operation of the resort.

## The Challenges of Ecotourism

Ecotourism requires the management and marketing of sensitive resources that often cross jurisdictional boundaries. As a result, many ecotourism projects face difficult land-use and regulatory challenges that arise from the conflicting goals of diverse agencies and private property owners.

Despite such challenges—and in part because of them—many state tourism officials regard ecotourism as a means for addressing existing problems of inter-agency conflict. It is not uncommon, for instance, for state park departments to face rising management costs as parks and other facilities experience increasing recreational use from activities that have become popular over the past few decades, among them mountain biking, wind surfing, and rock climbing.

While state tourism departments applaud such increased usage, agencies charged with management of natural resources—often understaffed and overwhelmed—may seek ways to discourage it.

By creating an ecotourism task force, states can bring together tourism and natural-resource officials and staff in a forum aimed at defining compatible objectives and exploring ways to share funding.

Ecotourism also poses property-rights challenges. Efforts to establish ecotourism sites may involve private landowners who hold negative impressions of tourism or government.

*"Most state tourism departments are involved in examining or actively promoting ecotourism."*

Traditionally, tourism and economic-development agencies have worked with developers, operators, and private-property owners who want to invest in—and profit from—business growth.

Such issues gripped the Greylock Glen project. In fact, it took more than three years of citizen participation before problems were resolved. Public participation allowed local residents to define what they wanted from the project and to settle differences between environmental groups and pro-growth groups. The result was a declaration of sustainable development adopted by the town council and a major zoning change to guide future growth.

Another issue that begs for resolution involves ecotourism's underlying principle of not over burdening an area's natural resources. Abiding by such a principle means someone must define resource-protection criteria. Setting such management standards involves, to some extent, deciding what people can, and cannot, do with their land.

This inherent challenge, along with long-term growth concerns, prompted British Columbia to create a province-wide Commission on Resources and Environment. The goals of the commission's land-use planning process are to (1) protect and restore the environment, and (2) secure a sound and prosperous economy. Defining tourism's potential impacts on key resources is one of the commission's tasks.

> *"Ecotourism requires the management and marketing of sensitive resources that often cross jurisdictional boundaries."*

Despite its honorable goals, the commission has run into significant landowner opposition as some people resist what they see as a province attempting to micro-manage private resources.

In whatever form it may take, ecotourism impacts must be defined. Such an effort will involve a new level of interaction among landowners, state and local resource-management agencies, and agencies intent on boosting ecotourism opportunities.

## The Ethics of Ecotourism

Ecotourism's ethics are laudable from a distance, but what happens when an operator's profits decline during economic recession? He or she may, in the interest of economic survival, drop environmental standards, increase the number and size of tourist groups, and otherwise violate ecotourism ethics. When that occurs, who will monitor the operators?

Since most ecotourism resources are either publicly owned or accessible to the public, government must assume some responsibility for ensuring ecotourism does not contribute to environmental degradation.

If the public owns the land, then local, state, or federal licensing requirements or management regulations provide the necessary oversight. In cases where the lands in question are privately owned, however, government responsibility and influence diminish. Outside of zoning or environmental-protection laws, resource management must depend on cooperative agreements among the ecotour operator, land owners, and government.

Such a management approach is in place in northern Maine, where most land is owned by paper companies and where whitewater rafting along the region's rivers is extensive.

Paper and rafting companies have worked together to strike an agreement that benefits both sides. Rafters pay fees and have access to the network of wilderness roads that the paper companies maintain. Meanwhile, the paper companies coordinate cutting practices to minimize adverse impacts to the landscape.

Greenville, Maine, is now moving ahead with an interpretive center that will present to the public the paper companies' and rafters' perspectives on cooperative land management.

## The Benefits of Ecotourism

Ecotourism—particularly on rural public lands—has become an increasingly popular economic-development tool because of its ability to satisfy environmentalists seeking to protect natural areas while soothing politicians eager to find economic-development solutions for economically depressed rural areas.

Its assets are appealing. Little public capital is required to develop and promote an ecotourism site on public lands. Public lands, in turn, can be used at little cost to tour operators. Entry-level jobs in the ecotourism market are often low skilled and thus fit the resumes of many rural workers. Meanwhile, rafting, mountain biking, cross-country skiing, hiking, and other eco-activities provide attractive entrepreneurial opportunities. Finally, ecotourism can allow the environment to be used, appreciated, and sometimes even improved.

## Potential Problems

While it's difficult to find fault with this line of thinking, flaws occur in expectations. "Ecotourism" is used to describe activities ranging from hiking the Appalachian Trail to participating in expensive guided tours to hard-to-reach ecosystems.

The former involves minimal daily expenditures and therefore minimal local economic impact. The latter involves paying for transportation, lodging, meals, and interpretive services and thus generates high economic impacts. Therefore, if areas cannot offer high-ticket experiences, ecotourism does not necessarily guarantee high economic returns.

In addition, there is the issue of resource protection and the resulting economic consequences. While technology and professional management can minimize the impacts of people on a resource, these ingredients are not always present.

> *"Growth in ecotourism will bring the needs of the environment to the forefront of the tourism agenda."*

For fragile ecological resources, there will be limits to the numbers of people who can visit a site before damage occurs. Meanwhile, in rural areas with average natural-resource attractions and limited capacities to manage visitation, ecotourism's potential must be viewed realistically. For these areas, ecotourism may offer only a modest increase in local revenues.

Communities hoping to build large-scale ecotourism industries must realize that visitors require accommodations, restaurants, and other entertainment diversions. Ecotourists may contribute to traffic and pollution, and they may disrupt the traditional tenor of country life.

## A New Standard

Ecotourism is a growing field, and some experts expect it to set a new standard for environmentally sensitive tourism that will affect the entire industry.

Though no one knows with certainty ecotourism's long-term consequences, it is safe to assume that growth in ecotourism will bring the needs of the environment to the forefront of the tourism agenda.

It will encourage travelers, as well as the tourist industry, to acknowledge the sensitive qualities of natural systems and to raise the public's environmental awareness. Coupled with its economic development potential, ecotourism offers benefits that all states should consider.

# Environmental Regulations Infringe on Property Rights

## by Sigfredo A. Cabrera

**About the author:** *Sigfredo A. Cabrera is the director of communications for the Pacific Legal Foundation in Sacramento, California, a nonprofit legal foundation that fights for individual rights and economic freedom.*

> The moment that idea is admitted into society that property is not as sacred as the Laws of God, and that there is not a force of law and public justice to protect it, anarchy and tyranny commence. —JOHN ADAMS

According to Black's Law Dictionary, the term *property* "embraces everything which is or may be the subject of ownership." It is the "unrestricted and exclusive right to a thing; the right to dispose of a thing in every legal way, to possess it, to use it, and to exclude everyone else from interfering with it." By definition, the term does not just apply to lumber companies, builders, ranchers, and farmers. If you own a home or business, you are a property owner. If you own a car, stocks, bonds, or an IRA, you are a property owner.

## Property Rights and Civil Rights

It is often overlooked (or perhaps ignored) that private property rights are included as civil rights guaranteed by the United States Constitution. The Fifth Amendment declares that "no person shall be . . . deprived of life, liberty or *property* without due process of law. . . ." That Amendment further states, "nor shall *private property* be taken for public use, without just compensation." And in the Fourteenth Amendment, local officials are forewarned, "nor shall any state deprive any person of life, liberty, or *property*."

Writing for the majority in 1994's landmark ruling in *Dolan v. City of Tigard*, Chief Justice William Rehnquist of the U.S. Supreme Court stated that property rights are as important a part of the Bill of Rights as freedom of speech and religion or the protection against unreasonable searches and seizures: "We see no reason why the Takings Clause of the Fifth Amendment, as much a part of the

Reprinted, by permission, from Sigfredo A. Cabrera, "Environmental Law Endangers Property Rights," *The Freeman*, August 1995.

Bill of Rights as the First Amendment or Fourth Amendment, should be relegated to the status of a poor relation."

All other civil and political rights—the right of basic freedom, religious worship, free speech, the right to vote—are vitally dependent on the right to own private property. "Let the people have property," said Noah Webster, "and they will have power—a power that will forever be exerted to prevent the restriction of the press, the abolition of trial by jury, or the abridgement of any other privilege."

## Property Rights Are Under Attack

History has taught painfully what hostility toward private property rights accomplishes. The social and economic travesty caused by over 70 years of Communist control of private property in the former Soviet Union is a lesson that should neither be forgotten nor repeated. But that lesson has not been heeded by those writing and enforcing modern environmental laws.

Like rust eating away metal until it crumbles, the erosion of property rights is a very slow and subtle process that can take not just months, but years, even generations—one instance, one case at a time. And nearly always, the erosion is not apparent. It is "behind the scenes"—not evident on the evening news or in the daily newspapers, but buried in thousands of pages of documents accumulated each year around the country in the corridors of government. Indeed, this country's fourth president, James Madison, stated in 1788: "I believe there are more instances of the abridgement of the freedom of the people by gradual and silent encroachments of those in power than by violent and sudden usurpations."

> *"It is often overlooked (or perhaps ignored) that private property rights are included as civil rights guaranteed by the United States Constitution."*

## The Ocie and Carey Mills' Case

On May 15, 1989, 58-year-old retiree Ocie Mills and his son Carey shocked the nation by becoming one of the first people to serve jail time for violating federal wetlands regulations. Their crime? Cleaning out a drainage ditch and putting clean sand on a parcel of land where Carey Mills planned to build a home. The Millses wanted to clean out the ditch to control mosquitos and to improve drainage. Although Ocie and Carey Mills had prior approval from the Florida Department of Environmental Regulation (DER), the U.S. Army Corps of Engineers (Corps) arrested them for filling in a "wetland" without a permit.

Believing the charges to be totally unfounded, Ocie did not hire an attorney, but defended himself and his son. "The charges were so incredibly trivial," he said, "I did not take them seriously and certainly didn't think that we could be in jeopardy of going to prison."

During their trial in Federal District Court, the judge refused to allow Ocie to present evidence confirming that the Millses' maintenance of the drainage ditch was allowed under Florida law and that DER officials authorized the placement of sand on his property. The judge also refused to allow DER employees to give their opinion that the property was not a wetland as defined by the Corps' regulations. Ultimately, the two men were each sentenced to 21 months in federal prison camp, were denied eligibility for parole, were each fined $5,000, and subsequently were ordered to restore the affected site within 90 days of their release.

## Further Rulings

After serving their time, the Millses were home with their family the day before Thanksgiving, 1990. But their ordeal would not be over. In March 1991, federal officials hauled the Millses back into court on charges that they failed to comply with the probation order to restore the property. After personally examining the property, U.S. District Judge Roger Vinson sided with the Millses and ruled that the "defendants have substantially complied with the site restoration plan." In his ruling he noted that the Corps' mandated "restoration" had left the lots "totally denuded and ugly" and that further "restoration" as required by the Corps would destroy the property's value.

In the spring of 1992, the Millses went back to the U.S. District Court to erase their convictions. But constrained by the present state of the law, the reluctant and sympathetic judge upheld their convictions. In his March 1993 ruling, Judge Vinson expressed astonishment of how the federal Clean Water Act had been interpreted in a manner "worthy of *Alice in Wonderland*" in which "a landowner who places clean fill dirt on . . . dry land may be imprisoned for . . . discharging pollutants into the navigable waters of the United States." The Eleventh Circuit Court of Appeals in Atlanta upheld their convictions on October 27, 1994; on May 15, 1995, the U.S. Supreme Court turned down their request for review. . . .

## A Dilemma in Oregon

In 1983, Tom Dodd and his wife, Doris, had put $33,000 of their life's savings into a 40-acre, scenic parcel in Hood River County, Oregon, overlooking beautiful Mt. Hood. A major factor in their decision to buy the lot was the prior assurances they received from local officials that building a home there was permitted. But a short time later, the zoning was changed. Under the new rules, they can use their property only for growing and harvesting lumber. A house is permitted only if absolutely necessary to accommodate a full-time forester on the property.

> *"The erosion of property rights is a very slow and subtle process that can take not just months, but years, even generations."*

Twenty-two acres of the property are covered by a type of soil that will not support forest vegetation. The combined value of the land as now zoned and the estimated proceeds from harvesting the few merchantable trees from the forested area would be less than $700! Moreover, according to a forest expert, harvesting trees on the parcel would damage watershed yields, wildlife habitat, aesthetic qualities, and the protection to neighboring properties from wind.

As retirees, the Dodds have no desire to engage in the forestry business, and they certainly do not wish to be forced into a losing business venture. And so the inescapable conclusion is that unless Tom and Doris are allowed to build their house, their property is useless to them. After exhausting every possible administrative avenue and failing in the Oregon court system, the Dodds have now taken their fight into the federal court system. A ruling from the Ninth Circuit Court of Appeals was expected in 1995. [On June 29, 1995, the appellate court remanded the Dodd's Fifth Amendment taking claim to the federal trial court. The county agreed to give the Dodds a land use permit, but the case has continued.]

## Lois Jemtegaard's Vacant Lot

Mrs. Lois Jemtegaard of Skamania County, Washington, owns a vacant 20-acre parcel that the county zoned for a single-family home. She would like to sell the parcel as a buildable lot so she would have money to repair her home, located on another parcel, that she says "is literally falling down around my ears." The proceeds would also help supplement the widow's retirement income.

The problem is that the parcel she wants to sell is considered to be a "resource" and "scenic" land under the Columbia River Gorge National Scenic Area Act. Under that federal law, the parcel may be used only for agriculture or timber operations.

> *"Government is supposed to protect private property— . . . not to destroy its economic value through overregulation."*

However, the property is not presently suitable for either of those uses.

Although Mrs. Jemtegaard holds formal title to the property, for all practical purposes she has lost any realistic use of it. Moreover, she has not received a nickel of compensation for the "taking" of her land for public benefit. Her parcel has lost its economic value as a buildable lot so long as the Columbia River Gorge Commission's decision disallowing a home remains in effect.

## The Effects of Regulation

These instances of environmental regulation gone amok in America represent only the tip of an ugly iceberg whose body is submerged and invisible to most of us. Many more "silent encroachments" can be found in the legal files of Pacific Legal Foundation, a nonprofit organization defending in court the property rights of the Millses, the Dodds, Mrs. Jemtegaard, and others like them.

We are witnessing a gradual decay in the basic principle that government is supposed to *protect* private property—not to take it away, not to impede reasonable use and enjoyment of it, and not to destroy its economic value through overregulation. It is critically important that citizens stay informed and communicate their concerns to their elected representatives about proposed or existing policies that are harmful to private property rights.

*"In courtroom battles involving land use and environmental protection, the interests of mainstream Americans are typically* **under-represented.***"*

Environmental laws are too often churned out with little or no regard for their costs or their consequences to human life, private property rights, and the free enterprise system. Under the federal Endangered Species Act, vast areas of land suitable for housing or other beneficial uses are being closed off to development, because of findings that the land is a current or potential habitat of some endangered or threatened animal, fish, or plant. Appalling as it may seem, the social, economic, or environmental benefits of proposed projects are deemed *irrelevant* by federal regulators who decide if a species should be protected. Human existence is simply disregarded in efforts to save certain species.

## Expensive Endangered Species

The Delhi Sands Flower-Loving Fly, whose lifespan is about 10 days, enjoys the same protected status as the American bald eagle, grizzly bear, and California condor. Swat this little creature and you could face a year in jail and up to $200,000 in fines! This obscure insect, which inhabits 700 scattered acres in San Bernardino County, California, now threatens to hinder needed economic development in the area. The detrimental effect of this kind of overzealous regulation is aptly illustrated in the following abstract of a report entitled, "Impacts of Mitigation for the Endangered Delhi Sands Flower-Loving Fly on the San Bernardino County Medical Center":

> The Endangered Species Act as applied to the construction of the San Bernardino County Medical Center resulted in an expenditure of $3,310,199 to mitigate for the presence of eight Delhi Sands Flower-Loving Flies. The effort as negotiated with the U.S. Fish and Wildlife Service and California Department of Fish and Game resulted in moving and redesigning the facility to provide 1.92 acres of protected habitat for eight flies believed to occupy the site. The effort mitigates only for species on site. Cost per fly amounted to $413,774.25 and resulted in a one year construction delay. This cost is equivalent to the average cost of treatment of 494 inpatients or 23,644 outpatients.

When fires swept Southern California in October 1994, the rural Winchester area of south Riverside County was hit particularly hard. Over 25,000 acres were charred and 29 homes destroyed. Many burned-out families in that area believe they could have saved their homes if only government officials had given them permission to create firebreaks around them. Brush fires can be kept away from homes by clearing out a strip of vegetation—a process called disking. Many of the victims of the Winchester fire have disked their property for years. But a few years ago officials from the U.S. Fish and Wildlife Service dissuaded them because doing so would disturb the burrows of the Stephens Kangaroo Rat, a rodent put on the federal endangered species list in 1988.

The Endangered Species Act either bans or strictly limits development on most of the 77,000 acres designated as "rat study" areas in Riverside County. Yshmael Garcia, a rancher who lost his home in the blaze, was quoted in the *Los Angeles Times:* "My home was destroyed by a bunch of bureaucrats in suits and so-called environmentalists who say animals are more important than people. I'm now homeless, and it all began with a little rat."

## The Fight for Property Rights

There is no shortage in this country of organizations dedicated to representing the interests of various species of animals or plants. Unfortunately, in courtroom battles involving land use and environmental protection, the interests of mainstream Americans are typically *under-represented.*

Every intrusive land-use or environmental regulation that is upheld in court results in the creation of a legal principle that acts like a building block upon which another antiproperty legal principle can be erected in yet another case. Years of bad precedent inevitably will result in a frail social and economic fabric that will not hold up to the wear of tyranny. That is why Americans must begin to stop the legal erosion of property rights, and restore this bulwark of our personal liberties.

# Environmental Regulations Are Necessary

by David Malin Roodman

**About the author:** *David Malin Roodman is a senior researcher at the Worldwatch Institute, an environmental research organization.*

Taxes and regulations are often cast as polar opposites. Regulatory approaches, born out of environmentalists' distrust of businesses, are the bad old policies, or so the fable runs. The best way to make sure firms cleaned up was to tell companies exactly how to do it. But now regulations have become burdensomely complex and perverse, wasting businesses' money and often failing to protect the environment as well. Tax and permit systems are the coming fashion. They can sweep away the tangle of rules, freeing business from its regulatory shackles.

Like most fables, this one contains some truth. Most of the bricks in the environmental policy edifice built during the last 30 years have been fired from the stuff of legal codes, not tax codes. And to be sure, governments have often overstretched regulation, and barely tapped the potential of market approaches. But the full truth is that the two approaches are best seen as complements, not rivals. Elegant in theory, tax and permit systems rarely work so neatly in practice. And less-pretty regulations have done much environmental good. Making the industrial economy operate efficiently within environmental limits will require synthesizing the two approaches, using the strengths of each to compensate for the weaknesses of the other.

## The Benefits of Regulation

Environmental regulations on the books have scored important successes. In Western Europe, for example, regulators can point to a 47 percent reduction in sulfur emissions between 1970 and 1993, due substantially to rules requiring scrubbers in coal plants. In the United States, tightened tailpipe emissions standards for new cars and light trucks made catalytic converters universal over the

Reprinted, by permission, from David Malin Roodman, "Getting Signals Right: Tax Reform to Protect the Environment and the Economy," *Worldwatch Paper 134*, Worldwatch Institute, May 1997.

same time span, cutting $NO_x$ emissions 6 percent, carbon monoxide 33 percent, and volatile organic compounds (smog ingredients) 54 percent, all despite a 44 percent increase in driving.

Moreover non-market approaches will continue to be essential to protecting the public interest. Laws—not market forces alone—are what will protect endangered species, manage nuclear waste, and ban pollutants that may be deemed unacceptable in any amount, such as DDT or dioxins. Waste incinerators, as long as they are built, are likely to be disproportionately located in poor and minority neighborhoods unless these communities have the legal means to protect themselves.

Contrary to popular belief, there is remarkably little evidence that regulations have seriously depressed the fortunes of industry, or that they have chased businesses into "pollution havens"—countries with lax environmental rules. For example, between 1970 and 1990, U.S. industries making and exporting the most pollution-intensive products such as paper and chemicals—all big spenders on regulation-required pollution control—fared better as a group in global competition than less-polluting industries, according to Robert Repetto, an economist at the World Resources Institute in Washington, D.C. A 1992 World Bank literature survey concluded that "the many empirical studies which have attempted to test these hypotheses [of regulatory harm] have shown no evidence to support them."

Moreover, debates over the costs of regulation have also often obscured their societal benefits, such as lower medical bills thanks to cleaner air. One study that has analyzed the benefits, also led by Repetto, found them to be substantial for some industries. Regulations were one reason the U.S. electric utility industry, for instance, spent 8 percent more for the amount of power sold in 1991 than it did in 1970. But once the economic benefits of less air pollution were factored in—an unorthodox but essential step in understanding the economics of regulation—this apparent productivity decline turned into an 8–15 percent productivity increase from the point of view of society as a whole. And more fundamentally, though less quantifiably, the regulations nudged the country closer to the ideal of the society that respects the right to a healthy environment.

> *"Laws—not market forces alone—are what will protect endangered species, manage nuclear waste, and ban pollutants."*

Regulations sometimes even benefit polluters. The prod of new pollution rules often stimulates companies to change and innovate, making them more, not less, competitive, argues Michael Porter of the Harvard Business School, who has studied dozens of examples. Searching for ways to cut resource waste, some managers have discovered ways to cut financial waste—or create new products. Executives at Rhone-Poulenc in Chalampe, France, discovered just such a connection when air pollution rules forced them to install $12 million worth of equipment to recover diacids, byproducts of nylon-making, rather than incinerate them. They now earn $3 million a year selling the chemicals as dye and tanning additives and coagulation agents, a handsome return on investment.

## An Alternative to Regulations

Nevertheless, regulations are increasingly being pushed beyond their limits. Because they often focus on means rather than ends, they tend to discourage innovation. And though they may work well when there is a front-runner solution (such as catalytic converters), they tend to break down in the face of complexity. A joint EPA-Amoco Corporation study documented one telling absurdity at the oil company's Yorktown, Virginia, refinery. Regulations required Amoco to spend $31 million on a wastewater treatment plant to stop airborne emissions of benzene, a carcinogen. Meanwhile, the rules failed to cover benzene emissions from a nearby loading dock—which could have been reduced as much for just $6 million. As one exasperated refinery official put it, "Give us a goal to meet rather than all the regulations.... That worked in the 1970's, when the pollution problems were much more visible and simpler. It's not working now."

The growing use of environmental tax and permit systems is one response to that plea. Whenever environmental goals can be expressed in a single number—how many tons of benzene should be permitted into an airshed each year, for example, or

*"Environmental problems sometimes defy measurement, making it difficult to apply quotas or taxes."*

how much water pumped from an aquifer—and whenever actual pollution or depletion rates can be estimated, market mechanisms offer an alternative. Such quantifiable problems include: urban smog; acid rain; overfishing; depletion and pollution of ground and surface water; and emissions of airborne toxics, ozone-depleting chemicals, and greenhouse gases. Market mechanisms can also supplement regulations in the more complex tasks of managing ecosystems and planning land use.

Thus tax and permit systems seem to offer all the benefits of regulations without the disadvantages. They allow governments to do what they do best—set targets for reducing environmental damage—while letting the market do what it does best—find the cheapest ways to get there. And many of the defenses of regulation—based on the often minimal harm to competitiveness, the economic benefits, the ability to stimulate innovation—apply at least as well to market-based approaches.

## The Market Approach Has Shortcomings

But taxes are not instant cures for environmental ailments. For instance, low-density zoning laws and heavy highway spending often give people little choice but to drive. To the extent that this happens (and to the extent that they already drive small cars), taxing gas does not so much discourage gas use as punish it. If governments of sprawl-afflicted countries such as the United States and Australia want gas tax hikes to work well, and fairly, they will need to give people better alternatives to driving. They will need to spend more on mass transit, as much of Western Europe already does, and rewrite zoning laws to foster neighborhoods that more intimately mix schools, homes, and shops. Together, these changes can lure people from behind the wheel and onto sidewalks, bike paths, or bus lines.

> *"Along with regulations, other government programs can work synergistically with environmental taxes."*

Another shortcoming of the pure market approach is that the environmental problems sometimes defy measurement, making it difficult to apply quotas or taxes. As a result, governments often have to tax rough proxies for actual pollution. Sweden, for example, taxes fertilizer sales rather than the amount of fertilizer that drains into surface and ground water, since that would be impractical to gauge. The exhaust spewing unmonitored from millions of cars provides another classic example. During annual car inspections, governments could perform odometer readings and emissions tests and use them to at least estimate a car's pollution over the past year, providing a rough base for a tax.

Unfortunately, the more policymakers latch onto what is easiest to measure rather than most relevant (like the man who looked for his keys near the lamppost because that was where he could see the sidewalk), the less effective taxes become. Often, the best idea is to call non-market approaches to the assistance of market approaches, as World Bank economist Gunnar Eskeland found during detailed studies of pollution control in Chile, Mexico, and Indonesia. In Mexico City, one of the world's most smog-burdened cities, administrative expense and corruption make it nearly impossible to institute taxes based on annual car inspections. The practical palliative, he concluded, would be to require catalytic converters in new cars. But while converters can dramatically lower emissions per kilometer driven, they do nothing to reduce driving. That is where the gas tax comes in.

## Combining the Approaches

Regulations can also lower other barriers to tax and permit effectiveness. Even with today's low energy prices, for instance, consumers and companies often miss opportunities to save money by investing in energy-efficient appliances. Evidently, they do not respond to market signals nearly as nimbly as economists would wish. Regulations can help make the decisions for them, by blocking off the most wasteful options. For example, efficiency standards adopted in the United States during the last 10 years on refrigerators, fluorescent lights, and other appliances have pushed the average efficiency of new models up sharply, and will eventually save households an average $250 a year. If governments taxed energy or carbon emissions, that would only make efficiency standards more valuable in shunting consumers away from the most energy-guzzling equipment.

Along with regulations, other government programs can work synergistically with environmental taxes. The taxes can even provide the funding for such efforts. The Swedish government, for example, credits the dramatic drop in fertilizer use during the 1980s both to the taxes on fertilizer sales and to education programs, funded by the new revenues, that raised farmers' awareness of the financial and environmental costs of overuse.

Environmental policymakers need to resist the temptation to throw out the old when they bring in the new. The problems at hand are grave, demanding hard-headed, practical decisionmaking. If regulations or other programs have something to offer, then they should be used, in place of or in addition to tax and permit systems. Market-based policies will work best if the signals they send are orchestrated with policies across the entire apparatus of government.

# Emissions Trading Does Not Protect the Environment

by Brian Tokar

**About the author:** *Brian Tokar is a faculty member at Goddard College and Institute for Social Ecology in Plainfield, Vermont. He is the author of* Earth for Sale: Reclaiming Ecology in the Age of Corporate Greenwash.

> . . . the IPUAIC was a creature of the smog, born of the need to give those working to produce the smog some hope of a life that was not all smog, and yet, at the same time, to celebrate its power.  —Italo Calvino, "Smog," 1958[1]

Since the beginnings of the Industrial Revolution, corporate managers have sought to obscure the social and environmental impacts of pollution. Like Calvino's bedraggled editor of a fictional trade journal improbably named *Purification*—the organ of an industry-sponsored Institute for the Purification of the Urban Atmosphere in Industrial Centers—corporate functionaries have obligingly stretched the truth to put the best possible face on their employers' destructive activities.

Extractive industries appropriated the language of conservation in the 1920s, elite think tanks took the initiative in environmental research in the 1950s, and corporations steadily increased their influence over the mainstream environmental movement during the 1980s. In the 1990s, however, these efforts have taken a bold new turn. Even as corporate lobbyists work tirelessly behind the scenes to dismantle decades' worth of environmental protections, a new generation of policy analysts and free market ideologues has successfully advanced the notion that corporations—and the capitalist market itself—are now the key to a cleaner environment. This updated version of corporate environmentalism has had a striking impact on national legislation, regulatory policies, and not surprisingly, the U.S. environmental movement.

Can the "free market" help to promote a clean environment? Can tropical

Reprinted, by permission, from Brian Tokar, *Earth for Sale: Reclaiming Ecology in the Age of Corporate Greenwash*, South End Press, 1997.

forests be protected in a manner favorable to the multinational banks? Do companies have "rights" to pollute that should be tradable as a commodity? Will corporations guide themselves toward becoming "environmentally responsible"?

A decade or two ago, few environmentalists would have taken these questions seriously. But thanks to a new wave of corporate public relations, such claims have risen to the top of the agenda of several national environmental groups. At the forefront of this initiative are think tanks such as the American Enterprise Institute and the Democratic Leadership Council's

> *"Corporate functionaries have obligingly stretched the truth to put the best possible face on their employers' destructive activities."*

Progressive Policy Institute, along with mainstream environmental groups such as the influential Environmental Defense Fund. Corporate and government officials still frequently denounce environmentalists as enemies of economic progress, but now these same officials would have the public simultaneously believe that they are the true environmentalists. The mid-1990s have seen a redoubled effort by the champions of the "free market" to co-opt the environmental movement's often not-well-defined political direction. Many mainstream environmentalists have willingly allowed themselves to be taken along for the ride, particularly as "free market" environmentalism has become an important cornerstone of the Clinton administration's environmental policies.

Glossy catalogs of "environmental products," television commercials featuring environmental themes, and appropriation of the language of environmentalism to support corporate initiatives are merely the most visible hallmarks of the greenwashing of U.S. corporations.[2] Many of the same companies that support the Republican Right's wholesale assault on environmental regulation have retained high-priced public relations firms to create an environmentally friendly public image. John Stauber and Sheldon Rampton, in their probing study of the public relations industry, sarcastically titled *Toxic Sludge Is Good for You,* estimate that corporate America spends $1 billion a year for this cynical mixture of anti-environmental lobbying and environmentally friendly imagemaking. Stauber and Rampton found that one of the leading anti-environmental PR firms, Edelman Public Relations, credited with creating the virulently anti-environmental Alliance for America, has even gotten one of its executive vice-presidents, Leslie Dach, elected to the board of the National Audubon Society.[3]

## The New Environmentalism

The new corporate environmentalism goes much farther than adopting environmental language, airing television commercials, promoting "environmental products," or infiltrating high-profile environmental groups. It represents a wholesale effort to recast environmental protection based on a model of commercial transactions within the capitalist marketplace. "A new environmental-ism has emerged," wrote economist Robert Stavins, who has been associated with both the Environmental Defense Fund and the Progressive Policy Institute, "that embraces . . . market-oriented environmental protection policies."[4] Stavins directed a pioneering effort known as Project 88, which brought together environmentalists, academics, and government officials with representatives of Chevron, Monsanto, ARCO, and other major corporations. Its goal, back in 1988, was to propose new environmental initiatives to the administration of President-elect George Bush, featuring market incentives as a supplement to regulation. Project 88 was careful to distance itself from those who advocated "putting a price on our environment, assigning dollar values to environmental amenities or auctioning public lands to the highest bidder,"[5] [in the words of Senators Timothy Wirth and John Heinz]. Despite its relatively cautious tone, however, Project 88 opened the door to a much more sweeping rejection of regulation in favor of so-called market mechanisms.

## George Bush's Amendments

George Bush staged a significant media coup when he announced a series of proposed amendments to the federal Clean Air Act during the summer of 1989. Knowing that any positive initiative for the environment was bound to win accolades from a press corps well trained to emphasize presidential rhetoric over substance, Bush's advisers cast the proposal as a striking departure from Ronald Reagan's rabid anti-environmentalism. The plan would reduce acid rain–causing sulfur emissions, put large cities on a timetable for smog reduction, encourage the use of alternative fuels, and increase the regulation of toxic chemicals. Its most unique and far-reaching proposal would for the first time establish into U.S. law the practice of allowing companies to buy and sell the "right" to pollute.

> *"The new corporate environmentalism . . . represents a wholesale effort to recast environmental protection based on a model of commercial transactions."*

The idea of marketable pollution rights was a cornerstone of Project 88, but it did not entirely originate with Stavins and his colleagues. The Environmental Protection Agency (EPA) had experimented with a limited program of "emissions trading" since 1974, for the benefit of corporations like Du Pont, Amoco, USX, and 3M. These were mostly individually brokered deals, in which the EPA would allow companies to offset pollution from new industrial facilities by reducing existing emissions elsewhere, or negotiating with another company to do so. This approach has been used in the Los Angeles area and other cities to broker measured reductions of particular pollutants. In a 1979 article in the *Harvard Law Review,* a Harvard law professor named Stephen

Breyer, now a justice on the U.S. Supreme Court, proposed a more ambitious system of "marketable rights to pollute" as a possible alternative to both taxes and regulation.[6]

## The Development of Pollution Trading

Under the Bush plan, which became law as part of the Clean Air Act reauthorization of 1990, companies that reduced emissions of sulfur dioxide or other pollutants in one location would receive credits redeemable against higher emissions elsewhere. These credits could then be sold at a profit to other companies that were not in compliance with emissions standards or wished to build new facilities. It was the first attempt to extend emissions trading to the national level, to establish allowances that could be traded freely as a commodity, and to codify such trading into law as the centerpiece of a major regulatory program. Defenders of the plan claimed that the ability to profit from pollution credits would better encourage companies to invest in new pollution control technologies than would a system of fixed standards. They predicted tremendous savings to the economy, as the most cost-effective pollution reductions would be implemented first, and more expensive ones could be postponed until new technologies became available. As pollution standards would be tightened over time, proponents argued, the credits would become more lucrative and everybody could reap higher profits while fighting pollution.

In some political circles, the idea of making pollution a tradable commodity raised considerable alarm. Would they someday be selling cancer bonds on the New York Stock Exchange, as *Village Voice* columnist James Ridgeway suggested in the aftermath of Bush's speech? Or will we follow the course envisioned by Todd Gitlin in a *New York Times* op-ed piece, where he projected that states would soon be assigned quotas for murder, rape, and armed robbery, and people would go out shopping for armed robbery credits at the end of the year so they can get their children more Christmas presents?[7]

Yet the "pollution rights" provisions now enshrined in the Clean Air Act sparked surprisingly little controversy in the mainstream environmental movement, and opponents such as the U.S. Public Interest Research Group were effectively silenced. The debate has, for the most part, been limited to technicalities, such as what kinds of pollution are sufficiently transferable from one place to another—re-

> *"A closer look at the scheme for nationwide emissions trading reveals a certain cleverness amid the underlying folly."*

ductions in emissions causing acid rain some distance away should be traded, some proponents argue, while local rises and falls in smog levels in different regions should not. The Environmental Defense Fund has since proposed trading programs in western water rights and offshore fishing allotments, while others are seeking to make federal mining and grazing permits tradable on the open market. The ensuing discussions have been extremely revealing of the increasing influence of pro-corporate ideology within the environmental movement.

## The Basics of Trading

A closer look at the scheme for nationwide emissions trading reveals a certain cleverness amid the underlying folly. For true believers in the invisible hand of the market, it may seem positively ingenious. Here is how it works: The Clean Air Act amendments were designed to halt the spread of acid rain by requiring a 50 percent reduction in the total sulfur dioxide emissions from fossil fuel–burning power plants by the year 2000. Power plants were targeted as the largest contributors to acid rain, and participation by non-utility industrial polluters remained optional. To achieve this goal, utilities were granted transferable allowances to emit sulfur dioxide in proportion to their current emissions. This would become the "free-market" alternative to mandating emissions reductions, taxing the worst polluters, or underwriting the wider use of scrubbers and other pollution controls.

> *"Once the EPA actually began auctioning pollution credits in 1993, virtually nothing went according to their projections."*

Any facility that continued to pollute more than its allocated share would then have to buy allowances from someone who pollutes less. Emissions allowances were expected to begin selling for around $500 per ton of sulfur dioxide, with a theoretical ceiling of $2,000 per ton, which is the legal penalty for violating the new rules. Companies that could reduce emissions for less than the cost of the credits would be able to sell them at a profit, while those that lagged behind would have to keep buying credits at a steadily rising price. Firms could choose to purchase credits on the open market rather than implement additional pollution controls. Thus, it is argued, market forces would assure that the most cost-effective means of reducing acid rain will be implemented first, saving billions of dollars in pollution control costs and stimulating the development of new technologies.

There were numerous loopholes to entice utilities to participate in the program. A portion of the total emissions allowances were set aside to actually facilitate the construction of new projects. These were auctioned off beginning in 1993, with annual auctions of new allowances to continue indefinitely. Utilities get additional allowances for implementing an approved conservation plan. There are also many pages of rules for extensions and substitutions. The plan essentially eliminated requirements for backup systems on smokestack scrubbers and then eased the rules for estimating how much pollution is emitted when monitoring systems fail. With reduced emissions now a marketable commodity, the range of possible abuses will continue to grow, as utilities have a greater incentive than ever to cheat on reporting what comes out of their stacks.[8] "It's a bit like playing Wall Street or the Chicago Commodity Exchange," said one official of the utility industry's research arm, the Electric Power Research Institute.[9]

The comparison with more traditional forms of commodity trading came full circle in 1991, when it was announced that the entire system for trading and auctioning emissions allowances would be administered by the Chicago Board of Trade. Long famous for its ever frantic markets in everything from grain futures and pork bellies to foreign currencies, the Board is responsible for selling and auctioning allowances, maintaining a computer bulletin board to match buyers and sellers, and even establishing a futures market, ostensibly to protect allowance holders against price fluctuations. "While a small, but significant, step toward the ultimate creation of cash and futures market trading in emission allowances, this represents a larger step toward applying free-market techniques to address societal problems," proclaimed Chicago Board of Trade Chairman William O'Connor in a January 1992 press release.[10]

## Problems with Emissions Trading

But once the EPA actually began auctioning pollution credits in 1993, virtually nothing went according to their projections. The first pollution credits sold for between $122 and $310, significantly less than the agency's estimated minimum price, and by 1996 successful bids at the EPA's annual auction of sulfur dioxide allowances averaged $68 per ton of emissions.[11] Many utilities preferred to go ahead with pollution control projects, such as the installation of new scrubbers, that were planned before the credits became available. Others switched to low-sulfur coal and increased their use of natural gas in order to meet their eventual targets of 50 percent reductions in sulfur emissions.

Many companies questioned the viability of financial instruments such as pollution allowances, while others, most notably the North Carolina–based utility Duke Power, are aggressively buying allowances. At the 1995 EPA auction, Duke Power alone bought 35 percent of the short-term "spot" allowances and 60 percent of the long-term allowances, which are redeemable in the years 2001 and 2002. *Forbes* magazine blamed low participation on "regulatory uncertainty": utilities were concerned that state regulators would not permit them to include the cost of sulfur dioxide allowances in their rate base and raise customers' electric bills accordingly.[12]

The outcome of the EPA's experiment in emissions trading also reveals the inherent inequalities of such a system. Seven companies, including five utilities and two brokerage firms, bought 97 percent of the short-term "spot" allowances for sulfur dioxide emissions that were auctioned in 1995 and 92 percent of the longer-term allowances. The remaining few percent were purchased by a wide variety of people and organizations, including some who sincerely wished to take pollution allowances out of circulation. Students at several law schools raised hundreds of dollars, and a group at the Glens Falls Middle School on

Long Island raised $3,171 to purchase twenty-one allowances, equivalent to twenty-one tons of sulfur dioxide emissions over the course of a year. Unfortunately, this represented less than a 0.1 percent of the allowances auctioned off in 1995. By the fall of 1996, nearly $50 million in allowances had traded hands, in both public and private transactions. The Glens Falls group raised $20,000 for their 1996 effort, and were joined by six other middle and high school groups and fourteen additional nonprofit organizations, each raising much smaller amounts. These well-meaning, but ultimately naive, attempts to fight pollution by "buying" a few tons of sulfur dioxide at a time offer a curious testament to the emerging faith in market "solutions" to political problems.[13]

## A Policy That Does Not Work

Where pollution credits have been traded, their effect has often run counter to the program's stated intentions. One of the first publicized deals was a sale of credits by the Long Island Lighting Company to an unidentified midwestern company, raising concerns that regions suffering from the effects of acid rain were selling "pollution rights" to companies in regions where most of the pollution that causes acid rain originates. One of the first companies to bid for additional credits, the Illinois Power Company, canceled construction of a $350 million scrubber system in Decatur, Illinois. "Our compliance plan is based almost totally on purchase of credits," an Illinois Power spokesperson told the *Wall Street Journal.*[14]

At least one company has tried to cash in on the confusion by assembling packages of "multi-year streams of pollution rights," allowing utilities to defer or supplant purchases of new pollution control technologies. "What a scrubber really is, is a decision to buy a 30-year stream of allowances," John B. Henry of Clean Air Capital Markets told the *New York Times* with impeccable capitalist logic. "If the price of allowances declines in future years," paraphrased the *Times,* "the scrubber would look like a bad buy."[15] Meanwhile, supporters of tradable allowances continue to spin improbable claims. For example, Environmental Defense Fund director Fred Krupp told a business-oriented environmental magazine in 1994, "When companies receive credit for getting rid of sulfur dioxide, they are suddenly eager to search for, find and implement . . . innovative and cheaper technologies."[16] Next to such obfuscations, the cynical candor of a John B. Henry seems almost refreshing.

> *"Where pollution credits have been traded, their effect has often run counter to the program's stated intentions."*

Other proponents are more realistic. "With a tradeable permit system, technological improvement will normally result in lower control costs and falling permit prices, rather than declining emissions levels," wrote Robert Stavins (formerly of EDF) and Bradley Whitehead (a Cleveland-based management consultant with ties to the Rockefeller Foundation) in a 1992 policy paper published by the Democratic Leadership Council's Progressive Policy Institute.[17] In contrast to environmentalists like Fred Krupp of EDF, who have to defend their devotion to the new gospel of market environmentalism, these consultants are quite ready to concede that a tradable permit system is not likely to reduce pollution. Stavins and Whitehead further acknowledge, albeit in a footnote to an appendix, that the system can quite easily be compromised by large companies' "strategic behavior." Control of 10 percent of the market, they suggest, might be enough to allow firms to engage in "price-setting behavior." To the rest of us, it should be clear that if pollution permits are like any other commodity that can be bought, sold, and traded, then the largest "players" will have substantial control over the entire "game." Emission trading thus becomes yet another way to assure that large corporate interests will remain free to threaten public health and ecological survival, often with the willing consent of official environmentalism.

## *Notes*

1. Italo Calvino, "Smog," in *The Watcher and Other Stories* (San Diego, CA: Harcourt, Brace Jovanovich, 1971), p. 117.

2. For a series of comprehensive case studies, see *The Greenpeace Book of Greenwash* (Washington, D.C., Greenpeace international, 1992). On the proliferation of false environmental claims in advertising, see John Holusha, "Some Smog in Pledges to Help Environment," *New York Times*, April 19, 1990; Barry Meier, "Environmental Doubts on 'Green' Ads," *New York Times*, August 11, 1990; and John Holusha, "Coming Clean on Products: Ecological Claims Faulted," *New York Times*, March 12, 1991.

3. John Stauber and Sheldon Rampton, *Toxic Sludge is Good for You: Lies, Damn Lies and the Publication Relations Industry* (Monroe, Maine: Common Courage Press, 1995), pp. 125, 140; also see Chapter One, note 25.

4. Robert N. Stavins, "Harnessing Market Forces to Protect the Environment," *Environment*, vol. 31, no. 1, January/February 1989, p. 5

5. Senators Timothy Wirth and John Heinz, "Foreword" to *Project 88: Harnessing Market Forces to Protect Our Environment* (Washington, D.C.: Senate Offices of Wirth and Heinz, 1988), p. vii.

6. Stephen Breyer, "Analyzing regulatory failure, mismatches, less restrictive alternatives and reform," *Harvard Law Review*, vol. 92, no. 3, January 1979, p. 597.

7. James Ridgeway, "Catching the Wave," *Village Voice*, June 27, 1989; Todd Gitlin, "Buying the Right to Pollute? What's Next?" reprinted in the *Earth Day Wall Street Action Handbook* (New York, 1990).

8. U.S. EPA, "Auctions, Direct Sales and Independent Power Producers' Written Guarantee Regulations," *Federal Register*, vol. 56, no. 242, December 17, 1991; "Acid Rain Allowance Allocations and Reserves," *Federal Register*, vol. 57, no. 130, July 7, 1992.

9. Ian Torrens, quoted in Leslie Lamarre, "Responding to the Clean Air Act Challenge," *EPRI Journal*, April/May 1991, p. 23.

10. "CBOT board endorses submission to EPA to administer clean air allowance auction," *News from the Chicago Board of Trade*, January 21, 1992.

11. James Dao, "Some Regions Fear the Price As Pollution Rights Are Sold," *New York Times*, February 6, 1993; Matthew Wald, "Acid Rain Pollution Credits are Not Enticing Utilities," *New York Times*, June 5, 1995. EPA auction data is from the Acid Rain Pro-

gram Home Page on the World Wide Web (http://www.epa.gov/docs/acidrain/ard-home.html).

12. Paul Klebnikov, "Pollution rights, wronged," *Forbes*, November 22, 1993, p. 128.

13. U.S. EPA, "EPA Allowance Auction Results: list of Bidders for Spot Auction," "Transaction Summary of the Allowance Trading System," both from the U.S. EPA World Wide Web site (see note 11).

14. Jeffrey Taylor, "CBOT Plan for Pollution Rights Market Is Encountering Plenty of Competition," *Wall Street Journal*, August 24, 1993.

15. Matthew Wald, "He doesn't call them dirty deals," *New York Times* Business section, May 13, 1992.

16. "Beyond Bumper Sticker Slogans: ECO Interviews Fred Krupp of EDF," NY: *ECO*, January 1994, p. 30.

17. Robert Stavins and Bradley Whitehead, "The Greening of America's Taxes," *Progressive Policy Institute Policy Report*, no. 13, February 1992.

# Overreliance on Green Products Cannot Protect the Environment

## by Nina Rao

**About the author:** *Nina Rao is a communications intern with Zero Population Growth, an organization that seeks a sustainable balance of resources, the environment, and population throughout the world.*

A pat on the back, a round of applause, please, for all the recycling Americans! We have been faithfully crushing our cans, rinsing our jars, and dragging our bins to the curb every week. Now, let's take a minute to examine the results of our efforts. American Paper and Forest published the 1995 recovery rate of paper at 45.1%, up 11.6% since 1990. The American Plastics Council reported that 22.4% of plastic bottles sold in 1995 were recycled as compared to 9.0% in 1990. According to the Steel Recycling Institute, the recycling rate of steel cans was at 55.9% in 1995, up a staggering 31.3% since 1990. The Aluminum Association informs us that 2,017 million pounds of aluminum cans were recovered in 1995, as compared to 1,934 million in 1990.

## More Consumption, More Recycling

*The Characterization of Municipal Solid Waste in the United States*, a report prepared by Franklin Associates, Ltd. for the Environmental Protection Agency (EPA), totals the waste statistics and, yes, the trend continues. Total waste materials recovered in 1990 weighed 32.9 million tons; in 1994, they weighed 49.3 million tons (1995 figures are not yet available), and 7% more of total waste generation was recovered. The environmental industry employed 140 thousand more people in 1995 than in 1990, and its revenues rose by $33.5 billion over that same period. Gross Domestic Product (GDP) was up by 1500 billion, and the weather was definitely better. We seem to have won on all fronts. Clearly, our efforts have not been in vain.

Reprinted, by permission, from Nina Rao, "The Oxymoron of Green Consumption," *ZPG Reporter*, March/April 1997.

But, as we all know, statistics can be slippery, so let's take an extra minute to examine them more closely. The recovery rate of paper, for example, is found by dividing the amount of paper recovered by the total amount of paper supplied. In 1990, 86.901 million tons of paper were supplied and 29.094 of those were recovered. In 1995, 96.097 million tons were supplied and 43.348 recovered. In other words, although the percentage of paper recycled has increased, the total consumption of paper has also increased—by over 9 million tons in five years. Similarly, sales of aluminum cans increased by over 14 billion cans and sales of plastic bottles increased by over 1 billion pounds. But wait! The population has grown since 1990, so of course the amount of resources used has grown proportionately. However, the EPA report also informs us that waste generation per person per day is up from 4.3 to 4.4 pounds, which means that growth in waste generation is outpacing growth in population. Basically, we are recycling more than ever and consuming more than ever.

> *"We are recycling more than ever and consuming more than ever."*

But we are consuming more wisely, you may argue. We now buy "bio-degradable" detergents in "33% less-packaging" and "environmental" tissues made with 20% post-consumer content. We look for the circle of arrows and the Green Seal. Money talks, and industry is responding to our concerns. Marketing Intelligence Service, Ltd (MI), a market research firm located in Naples, New York, estimates that in 1996, 12.1% of new products introduced at the retail level made environmental claims. Janet Mansfield of MI says, "Green products have moved into mainstream markets." And this claim is borne out by the very fact that we are comfortable using words like "bio-degradable" and "recyclable." Awareness has undoubtedly grown. We can now choose "green" banks and telephone companies. We can choose un-bleached paper and soy-based inks. Kodak recycles film canisters. McDonald's has a vegetarian burger. We have made a difference! Just look at how wisely we've consumed!

## The Consumption Paradox

And there lies the paradox: we are consumers living in a consumer society and espousing conservation ideals. In the four year period from 1990 to 1994, personal consumption expenditures rose by $331.9 billion. According to MI, in that same time period 33,672 new products were introduced at the retail level. Again, though the number of green products on the market is rising, the total number of products and the total amount of consumption is rising as well. The fundamental problem is not one of directing consumption into green avenues; that addresses the symptom. The problem is one of reducing consumption.

The message that consumption is good and necessary is built into the fabric of our society and comes from endless sources. Economists lecture on supply and demand and fluctuations in employment. Politicians inform us that there needs to be a free flow of money, that we need to keep the gears oiled in order for the economy to run smoothly. Advertisers seduce us with images of razors that shave closer than ever and digital can openers that will save us time and hassle. "If you don't have this, you don't know what you've been missing," they suggest, and a little voice inside us responds, "I really *do* need that eye gel made from egg protein and papaya that will make me look ten years younger overnight." The bottom line is this: buy, buy, buy. Do it for your happiness; do it for your country. Just do it.

## The Three R's

The environmentalists' three R's go against this philosophy entirely. "Reduce, Reuse, Recycle" they tell us. Reduction comes first because it addresses the problem at its root. Much of what we have we do not need or even use. This is a society of excess. One Kansas City organization, the Surplus Exchange, capitalizes on this fact by collecting surplus materials from offices and making them available to other organizations. The idea is simple: the less we consume, the less we impact our environment. But the idea of reducing consumption seems un-American. It seems almost traitorous, a willful boycott of the economic well-being of our country. Our very culture is based on consumption. We have more sales than holidays. We spend our weekends in malls. No museum is complete without a gift shop.

To reuse is the second step. The U.S. does have a reuse culture. Garage sales, flea markets, and antique auctions, all American institutions, are proof of this, but we are also obsessed with newness. We want the latest television and the latest VCR. We love shiny refrigerators and stream-lined cars, and since replacement is often cheaper than repair, we trade them in at alarming rates. In 1995 alone, Americans spent $606.4 billion on durable goods (defined by the Bureau of Economic Analysis as commodities that "have an average life span of at least 3 years"—things such as household appliances, furniture, cars, etc.), up $129.9 billion from 1990. A fundamental problem of the materialist society is that, by its nature, it devalues goods on a discrete level, even as it deifies them on a composite level. Since we can always buy another microwave, there is little incentive to

> *"A fundamental problem of the materialist society is that, by its nature, it devalues goods on a discrete level."*

take care of the one we already have. No one darns their socks anymore; there is no point. New socks are a dime a dozen.

The final step is to recycle, and here we have found the answer to our dilemma: through recycling we can assuage our environmental conscience without changing our consumption patterns. We no longer have to feel guilty for using a full sheet of paper for a one-line memo because, after all, we recycle the paper afterwards. We no longer have to worry about buying single-serving yogurts because we recycle the plastic containers. Recycling is not cost-less, however. Even re-melting aluminum, one of the most efficient materials to re-cycle at a 95% energy savings, is not free. Furthermore, it cannot solve the problem of solid waste since about one-fourth of waste materials are not recy-clable (things like kitty litter, food scraps, etc.). Dr. J. Winston Porter, president of the Waste Policy Center in Leesburg, Virginia, predicts that "notwithstanding the major recycling progress in America, we are not go-ing to see dramatically increased recycling rates in the foreseeable future." Re-cycling is not a sustainable solution. It should be a last-ditch effort in waste management, not, as it has become, the spearhead of the environmental move-ment. As Paul Hawken points out in *The Ecology of Commerce*, we cannot "save the environment by recycling our Coke cans and burrito foils." This solu-tion is too simplistic.

> *"We do not necessarily need to stop consuming—we need to start consuming differently."*

Hawken writes, "We must design a system . . . where doing good is like falling off a log, where the natural, everyday acts of work and life accumulate into a better world as a matter of course, not a matter of conscious altruism." It is not enough for us to recycle; it is not even enough for us to be green con-sumers; we must lead green lives. We cannot afford to become complacent.

## Wealth and Happiness

Economic theory assumes that human wants are dynamic, in other words, that they cannot be satisfied. Once we have food, we want shelter, and once we have shelter, we want a blender. This proposition brings some interesting questions to mind. If human wants are endless, then how can a new car make us happy? If human appetites are insatiable, then what level of contentment will consump-tion bring? In fact, studies show that material wealth, in absolute terms, has lit-tle bearing on happiness levels. In *How Much Is Enough?*, Alan Durning writes, "Regular surveys by the National Opinion Research Center of the University of Chicago reveal that no more Americans report they are 'very happy' now than in 1957. The 'very happy' share of the population has fluctuated around one third since the mid-fifties, despite near-doublings in both gross national product and personal consumption expenditures per capita." Likewise, studies show that people in poorer countries are no more likely to be unhappy than people in richer countries.

What does seem to matter is *relative* wealth. People identify their needs by what they see around them. If the Smiths have a new super-deluxe computer, the Browns down the street will judge their level of well-being on the ability of owning one too. The Patels in India, however, will not judge their well-being by this computer (the absolute measure), but rather by the typewriter that the Raos next door have (the relative measure). The ideal answer would be to stop this cycle by not consuming, then the Brown family would never again feel the need to live up to the standards of the Smiths.

Very few of us are willing to forego material pleasures entirely and live as monks, and it is not required in order to affect change. "A rose is a rose is a rose is a rose," wrote Gertrude Stein. But that equation cannot be applied to consumption. One sort of consumption differs widely from another. We do not necessarily need to stop consuming—we need to start consuming differently. For example, despite seeming simple, buying a shirt is a global transaction. Raw materials are shipped from one continent to another for processing, manufacturing, and assembly. Simply the label, "Made in Spain" combines Indian textiles, Chinese inks, Indonesian labor, and German looms.

However, material goods consumption is only one form of consumption. Instead of buying things, we can buy experiences. Instead of treating ourselves to another bottle of perfume, we can treat ourselves to a massage. Massages involve no packaging or shipping and very little assembly. Instead of going to the mall, we can go to the theater, or—a truly revolutionary idea—instead of treating ourselves to more money, we can treat ourselves to more leisure time. There are choices to be made, and we, in the double role of consumer and producer, need to realize that we have freedom within our economic system. We must redefine—or maybe remember—what gives us pleasure. By choosing to work fewer hours (the average American work week is longer than in any other developed country), we can instead choose to spend more time with our families, pets, and hobbies. We must realize that the market does not control us. We control the market.

# The Free-Market System Harms the Environment

## by Vandana Shiva

**About the author:** *Vandana Shiva is the director of the Research Foundation for Science, Technology, and Natural Resource Policy in New Delhi, India.*

In the *Vishnu Purana* [a sacred Hindu text], the world is destroyed and recreated by the cosmic being when human values fail to maintain nature and society. Vishnu, the Creator, assumes the character of Rudra or Shiva, the destroyer, and descends to reunite all his creatures with himself. He enters into the seven rays of the sun and drinks up all the waters of the Earth, leaving the seas and the springs dry.

The reduction of all value to wealth and the exclusion of compassion and care from human relationships are among the factors that cause this dissolution. As the *Vishnu Purana* puts it: 'The minds of men will be wholly occupied in acquiring wealth, and wealth will be spent solely on selfish gratification. Men will fix their desires upon riches, even though dishonestly acquired. No man will part with the smallest fraction of the smallest coin, though entreated by a friend. The people will be almost always in dread of dearth and apprehensive of scarcity'.

## A False Belief in Monetary Values

The links between greed, scarcity and destruction that this story brings out are at the heart of the ecological crisis. The reduction of all value to monetary value is an important aspect of the crisis of scarcity generated by the process of increasing affluence.

It is often said that the roots of environmental destruction lie in treating natural resources as 'free' and not giving them 'value'. Most discussions in the dominant paradigm assume that monetary, commercial or market value is the only way of measuring or valuing the environment. It is falsely assumed that value can be reduced to price.

However, the market is not the only source of values, and monetary values are not the only ones. Spiritual values treat certain resources and ecosystems as sa-

Reprinted from Vandana Shiva, "Values Beyond Price," *Our Planet*, vol. 8, no. 2, 1996.

cred—there are also such social values as those associated with common property resources. In both cases, resources have no price—but a very high value. In fact, it is precisely because their value is high that these resources are not left to the market but are taken beyond the domain of monetary value so as to protect and conserve them.

The proposal to solve the ecological crisis by giving market values to all resources is like offering the disease as the cure. The reduction of all value to commercial value, and the removal of all spiritual, ecological, cultural and social limits to exploitation—the shift that took place at the time of industrialization—is central to the ecological crisis.

## Redefining Resources

This shift is reflected in the change in the meaning of the term 'resource', which originally implied life. Its root is the Latin verb, *surgere,* evoking the image of a spring continually rising from the ground. Like a spring, a 'resource' rises again and again, even if it has been repeatedly used and consumed. The word highlighted nature's power of self-regeneration and her prodigious creativity. Moreover, it implied an ancient idea about the relationship between humans and nature—that the Earth bestows gifts on humans who, in turn, are well advised not to suffocate her generosity. In early modern times, 'resources' therefore suggested reciprocity along with regeneration.

> *"The reduction of all value to commercial value . . . is central to the ecological crisis."*

With the advent of industrialism and colonialism, 'natural resources' became the parts of nature required as inputs for industrial production and colonial trade. In 1870 John Yeates in his *Natural History of Commerce* offered the first definition of this new meaning: 'In speaking of the natural resources of any country, we refer to the ore in the mine, the stone unquarried (etc).'

By this view, nature has been stripped of her creative power and turned into a container for raw materials waiting to be transformed into inputs for commodity production. Resources are merely any materials or conditions existing in nature which may have potential for 'economic exploitation'. Without the capacity of regeneration, the attitude of reciprocity has also lost ground: it is now simply human inventiveness and industry which 'impart value to nature'. Natural resources must be developed and nature will only find her destiny once capital and technology have been brought in. Nature, whose real nature it is to rise again, was transformed by this originally Western world view into dead and manipulatable matter—its capacity to renew and grow denied.

## More than One Economy

The market economy is only one of the world's economies—in addition, there is nature's economy of life-support processes and people's economy in which our sustenance is provided and our needs are met. Nature's economy is the most basic, both in that it is the base of the people's and market economies, and because it has the highest priority to, and claim on, natural resources. However, development and economic growth treat the market economy as the primary one, and either neglect the others or treat them as marginal and secondary.

Capital accumulation does lead to financial growth. but it erodes the natural resource base of all three economies. The result is a high level of ecological instability. The anarchy of growth and the ideology of development based on it are the prime reasons underlying the ecological crises and destruction of natural resources. In order to resolve ecological conflicts and regenerate nature these economies must be given their due place in the stable foundation of a healthy nature.

Commodification of resources must be replaced by the recovery of commons. This involves the recovery of the domains of nature's economy and the sustenance economy, which, in turn, involves the recovery of the value of nature in its spiritual, ecological and social dimensions.

## The Myths of Market Economics

The dominant model of environmental economics promoted by the World Bank and major economic powers attempts further to reduce nature's economy and the sustenance economy to the market economy. Preoccupation with 'getting the prices right' can lead to a blindness to the fact that the market usually gets the values of justice and sustainability wrong.

The marketization of common resources is based on myths. The first is the equivalence of 'value' and 'price'. Resources—such as sacred forests and rivers—often have very high value while having no price. The second is that common property resources tend to degrade. Privatization is frequently prescribed for solving 'problems' caused by overusing resources under open access and common property. But it is based on the tradeability of private property, while commons are based on the inalienability of shared rights derived from use. The assumption that alienability is more conducive to conservation is derived from the false association of price with value.

> *"Resources—such as sacred forests and rivers—often have very high value while having no price."*

It has been argued that landowners have little incentive to invest in long-term measures such as soil conservation if they do not have the right to sell or transfer their land, and thus cannot realize the value of any improvements. This is patently false, since the best examples of soil conservation—such as in the hill-terraces of the Himalaya—have been realized for precisely the opposite reasons. Communities who are not threatened by alienation of resources and their benefits have the long-term possibility and interest to conserve them.

194

## The Costs of Degradation

The dominant paradigm of environmental economics fails to internalize the costs of resource degradation socially and ecologically. Social internalization would imply that those responsible for environmental degradation should bear the costs of it.

Turning commons into commodities is a necessary part of environmental economics in the market paradigm. But it does not stop environmental degradation because the economically powerful do not mind paying a higher price for a resource. Other people bear the costs both of the scarcity of a declining resource, to which the rich can continue to have access, and of related scarcities and pollution caused by overexploitation. These ecological costs are not considered in the reductionist model of market internalization.

A genuine internalization would have to include values beyond those of the market, values that put limits on overexploitation. Given the vast gulf between the rich and poor, market prices, no matter how high they rise, will not introduce limits to exploitation. They will therefore not restrict resource exploitation within ecological limits, but will instead allow resource degradation to continue while aggravating poverty and injustice.

Economic growth takes place through the overexploitation of natural resources, creating a scarcity of them in both nature's economy and the survival economy. Nature shrinks as capital. The growth of the market cannot solve the very crisis it creates. Furthermore, while natural resources can be converted into cash, cash cannot be converted into nature's ecological processes. Those who offer market solutions to the ecological crisis limit themselves to the market, and look for substitutes to the commercial function of natural resources as commodities and raw material. However, in nature's economy, the currency is not money, it is life.

This neglect of the role of natural resources in ecological processes and in people's sustenance economy—and the diversion and destruction of these resources for commodity production and capital accumulation—are the main reasons for both the ecological crisis and the crisis of survival in the developing world. The solution seems to lie in giving local communities control over local resources so that they have the right and responsibility to rebuild nature's economy and, through it, their sustenance. Only this will ensure greater distributive justice, participation and sustainability.

# Bibliography

**Books**

Diane Ackerman — *The Rarest of the Rare: Vanishing Animals, Timeless Worlds.* New York: Random House, 1995.

Terry L. Anderson and Donald R. Leal — *Enviro-Capitalists: Doing Good While Doing Well.* Lanham, MD: Rowman and Littlefield, 1997.

John A. Baden, ed. — *Environmental Gore: A Constructive Response to Earth in the Balance.* San Francisco: Pacific Research Institute for Public Policy, 1994.

Ronald Bailey, ed. — *The True State of the Planet.* New York: Free Press, 1995.

Sharon Beder — *Global Spin: The Corporate Assault on Environmentalism.* White River Junction, VT: Chelsea Green, 1998.

James DeLong — *Property Matters: How Property Rights Are Under Assault—and Why You Should Care.* New York: Free Press, 1997.

Mark Dowie — *Losing Ground: American Environmentalism at the Close of the Twentieth Century.* Cambridge, MA: MIT Press, 1995.

Gregg Easterbrook — *A Moment on the Earth: The Coming Age of Environmental Optimism.* New York: Viking, 1995.

Paul R. Ehrlich and Anne H. Ehrlich — *Betrayal of Science and Reason: How Anti-Environmental Rhetoric Threatens Our Future.* Washington, DC: Island Press, 1996.

Paul R. Ehrlich, Anne H. Ehrlich, and Gretchen C. Daily — *The Stork and the Plow: The Equity Answer to the Human Dilemma.* New York: G.P. Putnam's Sons, 1995.

Ross Gelbspan — *The Heat Is On: The High Stakes Battle over Earth's Threatened Climate.* Reading, MA: Addison-Wesley, 1997.

Lindsey Grant — *Juggernaut: Growth on a Finite Planet.* Santa Ana, CA: Seven Locks Press, 1996.

David Helvarg — *The War Against the Greens: The "Wise-Use" Movement, the New Right, and Anti-Environmental Violence.* San Francisco: Sierra Club Books, 1994.

J. Robert Hunter — *Simple Things Won't Save the Earth.* Austin: University of Texas Press, 1997.

# Bibliography

| | |
|---|---|
| Wallace Kaufman | *No Turning Back: Dismantling the Fantasies of Environmental Thinking.* New York: BasicBooks, 1994. |
| Jane Holtz Kay | *Asphalt Nation: How the Automobile Took over America and How We Can Take It Back.* New York: Crown, 1997. |
| Richard E. Leakey and Roger Lewin | *The Sixth Extinction: Patterns of Life and the Future of Humankind.* New York: Doubleday, 1995. |
| Steve Lerner | *Eco-Pioneers: Practical Visionaries Solving Today's Environmental Problems.* Cambridge, MA: MIT Press, 1997. |
| Charles C. Mann and Mark L. Plummer | *Noah's Choice: The Future of Endangered Species.* New York: Knopf, 1995. |
| Curtis Moore and Alan Miller | *Green Gold: Japan, Germany, the United States, and the Race for Environmental Technology.* Boston: Beacon Press, 1994. |
| William Perry Pendley | *War on the West: Government Tyranny on America's Great Frontier.* Washington, DC: Regnery, 1995. |
| Michael Jay Polonsky and Alma T. Mintu-Wimsatt, eds. | *Environmental Marketing: Strategies, Practice, Theory, and Research.* Binghamton, NY: Haworth Press, 1995. |
| Charles T. Rubin | *The Green Crusade: Rethinking the Roots of Environmentalism.* New York: Free Press, 1994. |
| Michael Shnayerson | *The Car That Could: The Inside Story of GM's Revolutionary Electric Vehicle.* New York: Random House, 1996. |
| Julian L. Simon | *The Ultimate Resource 2.* Princeton, NJ: Princeton University Press, 1996. |
| Brian Tokar | *Earth for Sale: Reclaiming Ecology in the Age of Corporate Greenwash.* Boston: South End Press, 1997. |
| Bruce Yandle | *Land Rights: The 1990s' Property Rights Rebellion.* Lanham, MD: Rowman and Littlefield, 1995. |

## Periodicals

| | |
|---|---|
| Tom Arrandale | "Conservation by Consensus," *Governing,* July 1997. |
| Dennis T. Avery | "Global Warming—Boon for Mankind?" *American Outlook,* Spring 1998. Available from the Hudson Institute, Herman Kahn Center, 5395 Emerson Way, PO Box 26-919, Indianapolis, IN 46226. |
| Bruce Babbitt | "Stewards of Creation," *Christian Century,* May 8, 1996. |
| Jerry Brown | "Apocalypse Soon," *Spin,* December 1996. Available from 6 W. 18th St., New York, NY 10011. |
| Mary H. Cooper | "New Air Quality Standards" *CQ Researcher,* March 7, 1997. Available from 1414 22nd St. NW, Washington, DC 20037. |
| *Dollars and Sense* | Special Section: The Environment, March/April 1996. |

# Conserving the Environment

Nicholas Eberstadt — "The Magicians of Kyoto: Global Environmentalists and Their Superstitions," *Weekly Standard,* December 22, 1997. Available from PO Box 96153, Washington, DC 20090-6153.

Timothy Egan — "Look Who's Hugging Trees Now," *New York Times Magazine,* July 7, 1996.

Christopher Flavin and Odil Tunali — "Getting Warmer: Looking for a Way out of the Climate Impasse," *World Watch,* March/April 1995.

*Issues and Controversies On File* — "Endangered Species Act," June 13, 1997. Available from Facts On File News Service, 11 Penn Plaza, New York, NY 10001-2006.

*Issues and Controversies On File* — "Water Quality," February 9, 1996.

Jerry J. Jasinowski — "Business Is America's Leading Environmentalist," *Christian Science Monitor,* April 21, 1995.

Robert M. Lilienfeld and William L. Rathje — "Six Enviro-Myths," *New York Times,* January 21, 1995.

Craig McCormack and Jane S. Shaw — "Emissions Trading: Clearing the Air," *PERC Reports,* June 1996. Available from 502 S. 19th Ave., Suite 211, Bozeman, MT 59715.

Roger Meiners and Bruce Yandle — "Get the Government out of Environmental Control," *USA Today,* May 1996.

Richard D. Morgenstern — "Environmental Taxes: Is There a Double Dividend?" *Environment,* April 1996.

Will Nixon — "Too Little, Too Late," *In These Times,* July 8–21, 1996.

F. Sherwood Rowland — "Change of Atmosphere," *Our Planet,* vol. 9, no. 2. Available from United Nations Environment Programme, PO Box 30552, Nairobi, Kenya.

Barbara Ruben — "Getting the Wrong Ideas," *Environmental Action,* Spring 1995.

Lynn Scarlett — "Smogged Down," *Reason,* December 1996.

Richard L. Stroup — "Making Endangered Species Friends Instead of Enemies," *American Enterprise,* September/October 1995.

Richard L. Stroup and Jane S. Shaw — "An Environment Without Property Rights," *Freeman,* February 1997. Available from the Foundation for Economic Education, Inc., Irvington-on-Hudson, NY 10533.

Laura Tangley — "A New Brief for Nature," U.S. *News & World Report,* October 27, 1997.

Arthur Weissman — "Greener Market Place Means Cleaner World," *Forum for Applied Research and Public Policy,* Spring 1997.

# Organizations to Contact

The editors have compiled the following list of organizations concerned with the issues debated in this book. The descriptions are derived from materials provided by the organizations. All have publications or information available for interested readers. The list was compiled on the date of publication of the present volume; the information provided here may change. Be aware that many organizations take several weeks or longer to respond to inquiries, so allow as much time as possible.

**Cato Institute**
1000 Massachusetts Ave. NW, Washington, DC 20001-5403
(202) 842-0200 • fax: (202) 842-3490
e-mail: cato@cato.org • website: http://www.cato.org

The Cato Institute is a libertarian public policy research foundation dedicated to limiting the role of government and protecting individual liberties. The institute publishes the quarterly magazine *Regulation,* the bimonthly *Cato Policy Report,* and numerous books, including *Through Green-Colored Glasses: Environmentalism Reconsidered* and *Climate of Fear: Why We Shouldn't Worry About Global Warming.*

**Competitive Enterprise Institute (CEI)**
1001 Connecticut Ave. NW, Suite 1250, Washington, DC 20036
(202) 331-1010 • fax: (202) 331-0640
e-mail: info@cei.org • website: http://www.cei.org

CEI encourages the use of the free market and private property rights to protect the environment. It advocates removing governmental regulatory barriers and establishing a system in which the private sector would be responsible for the environment. CEI's publications include the monthly newsletter *CEI Update* and editorials in its On Point series, such as *Property Owners Deserve Equal Access to Justice.*

**Environment Canada**
10 Wellington St., Hull, Quebec, CANADA, K1A 0H3
(819) 997-2800
website: http://www.ec.gc.ca

Environment Canada is a department of the Canadian government whose goal is to achieve sustainable development in Canada through environmental protection and conservation. It publishes reports and fact sheets on a variety of environmental issues.

**Environmental Defense Fund**
257 Park Ave. South, New York, NY 10010
(212) 505-2100 • fax: (212) 505-0892
website: http://www.edf.org

The fund is a public interest organization of lawyers, scientists, and economists dedicated to the protection and improvement of environmental quality and public health. It publishes brochures, fact sheets, and the bimonthly *EDF Letter.*

**Foundation for Research on Economics and the Environment (FREE)**
945 Technology Blvd., Suite 101F, Bozeman, MT 59718
(406) 585-1776 • fax: (406) 585-3000
e-mail: free@mcn.net • website: http://www.free-eco.org

FREE is a research and education foundation committed to freedom, environmental quality, and economic progress. It works to reform environmental policy by using the principles of private property rights, the free market, and the rule of law. FREE publishes the quarterly newsletter *FREE Perspectives on Economics and the Environment* and produces a biweekly syndicated op-ed column.

**The Heritage Foundation**
214 Massachusetts Ave. NE, Washington, DC 20002
(800) 544-4843 • (202) 546-4400 • fax: (202) 544-2260
e-mail: pubs@heritage.org • website: http://www.heritage.org

The Heritage Foundation is a conservative think tank that supports the principles of free enterprise and limited government in environmental matters. Its many publications include the following position papers: "Can No One Stop the EPA?" "How to Talk About Property Rights: Why Protecting Property Rights Benefits All Americans," and "How to Help the Environment Without Destroying Jobs."

**National Audubon Society**
700 Broadway, New York, NY 10003
(212) 979-3000 • fax: (212) 979-3188
e-mail: webmaster@list.audubon.org • website: http://www.audubon.org

The society seeks to conserve and restore natural ecosystems, focusing on birds and other wildlife for the benefit of humanity and the earth's biological diversity. It publishes *Audubon* magazine and the *WatchList,* which identifies North American bird species that are at risk of becoming endangered.

**Natural Resources Defense Council (NRDC)**
40 W. 20th St., New York, NY 10011
(212) 727-2700
e-mail: nrdcinfo@nrdc.org • website: http://www.nrdc.org

NRDC is an environmental group composed of lawyers and scientists who conduct research, work to educate the public, and lobby and litigate for environmental issues. The council publishes the quarterly *Amicus Journal* as well as many books, pamphlets, brochures, and reports, many of which are available on its website.

**Negative Population Growth, Inc. (NPG)**
1608 20th St. NW, Suite 200, Washington, DC 20009
(202) 667-8950 • fax: (202) 667-8953
e-mail: npg@npg.org • website: http://npg.org

NPG works to educate the American public and political leaders about the detrimental effects of overpopulation on our environment and quality of life. NPG advocates a smaller, more sustainable U.S. population accomplished through voluntary incentives for smaller families and limits on immigration. NPG publishes position papers such as "Why We Need a Smaller U.S. Population and How We Can Achieve It" and "Immigration and U.S. Population Growth: An Environmental Perspective."

**Political Economy Research Center (PERC)**
502 S. 19th Ave., Suite 211, Bozeman, MT 59718-6872
(406) 587-9591 • fax: (406) 586-7555
e-mail: perc@perc.org • website: http://www.perc.org

PERC is a research and education foundation that focuses primarily on environmental and natural resource issues. It emphasizes the advantages of free markets and the importance of private property rights in environmental protection. PERC's publications include the monthly *PERC Reports* and papers in the PERC Policy Series such as "The Common Law: How It Protects the Environment."

**Rainforest Action Network (RAN)**
221 Pine St., Suite 500, San Francisco, CA 94104
(415) 398-4404 • fax: (415) 398-2732
e-mail: rainforest@ran.org • website: http://www.ran.org

RAN works to preserve the world's rain forests through activism addressing the logging and importation of tropical timber, cattle ranching in rain forests, and the rights of indigenous rain-forest peoples. It also seeks to educate the public about the environmental effects of tropical hardwood logging. RAN's publications include the monthly *Action Report* and the semiannual *World Rainforest Report.*

**Sierra Club**
85 Second St., 2nd Fl., San Francisco, CA 94105-3441
(415) 977-5500 • fax: (415) 977-5799
e-mail: information@sierraclub.org • website: http://www.sierraclub.org

The Sierra Club is a nonprofit public interest organization that promotes conservation of the natural environment by influencing public policy decisions—legislative, administrative, legal, and electoral. It publishes *Sierra* magazine as well as books on the environment.

**U.S. Environmental Protection Agency (EPA)**
401 M St. SW, Washington, DC 20460
(202) 260-2090
website: http://www.epa.gov

The EPA is the government agency charged with protecting human health and safeguarding the natural environment. It works to protect Americans from environmental health risks, enforce federal environmental regulations, and ensure that environmental protection is an integral consideration in U.S. policy. The EPA publishes many reports, fact sheets, and educational materials.

**U.S. Fish and Wildlife Service**
1250 25th St. NW, Washington, DC 20037
(202) 293-4800
website: http://www.fws.gov

The U.S. Fish and Wildlife Service is a network of regional offices, national wildlife refuges, research and development centers, national fish hatcheries, and wildlife law-enforcement agents. The service's primary goal is to conserve, protect, and enhance fish and wildlife and their habitats. It publishes an endangered species list as well as fact sheets, pamphlets, and information on the Endangered Species Act.

**Worldwatch Institute**
1776 Massachusetts Ave. NW, Washington, DC 20036-1904
(202) 452-1999 • fax: (202) 296-7365
e-mail: worldwatch@worldwatch.org • website: http://www.worldwatch.org

Worldwatch is a research organization that analyzes and calls attention to global problems, including environmental concerns such as the loss of cropland, forests, habitat, species, and water supplies. It compiles the annual *State of the World* and *Vital Signs* anthologies and publishes the bimonthly *Worldwatch* magazine as well as papers in the Environmental Alert series, such as *Fighting for Survival: Environmental Decline, Social Conflict, and the New Age of Insecurity.*

# Index

# Index